4.00

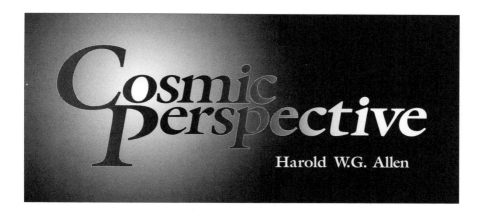

Cosmic Perspective

Harold W.G. Allen

Sunstar
PUBLISHING LTD.

Cosmic Perspective
by *Harold W.G. Allen*

© United States Copyright, 1998
Sunstar Publishing, Ltd.
116 North Court Street
Fairfield, Iowa 52556

All rights reserved. No part of this book may be reproduced or utilized in any form by any means, electronic or mechanical, including photocopying or recording, or by any information storage and retrieval system, without permission in writing from the publisher.

Cover Design: Theresa Cross
Cover Photo Courtesy of: Anglo-Australian Observatory
Photographer: David Malin, 1980
Editor: Sharon A. Dunn
Text Layout & Design: Sharon A. Dunn

LCCN: 97-066215
ISBN: 1-887472-21-5
Printed in the U.S.A.

Readers interested in obtaining further information on the subject matter of this book are invited to correspond with:

*The Secretary, Sunstar Publishing, Ltd.
116 North Court Street, Fairfield, Iowa 52556*

For more Sunstar Books, visit: *http://www.newagepage.com*

Dedication

It is a feature of our earthly existence that one may strive a lifetime in a vain quest for knowledge, while yet another may be so fortunate as to inherit vital information in the form of history and tradition. Scientific advances of the present generation alone have given us much precious insight that was denied to many inquisitive minds of the past. Only in a limited sense, therefore, may this work be considered a personal interpretation of our mysterious universe. Indeed, it was made possible solely through the contributions of many others, some of whom received not even the satisfaction of knowing the real worth of their discoveries.

Accordingly, it is to the memory of those who have sought in vain, to those that today and in the future are genuine in their search for Truth, that this book is hereby dedicated.

Table of Contents

Part One:
Science and Philosophy

Chapter 1. Evolution and Reincarnation 3
    The Procession of Life 4
    The Evolution of Man 11
    The Mechanism of Evolution 15
    The Pyramid Concept of Reincarnation 21

Chapter 2. Our Physical Universe 27
    The Enigma of Matter 28
    Solar Systems 30
    The Stars 33
    Degenerate Stars 35
    The Milky Way 38
    Galaxies of Stars 41
    The Expanding Universe 42
    Quasars 45
    The Creation Impasse 47

Chapter 3. Cosmological Theories 51
    Gravitational Attraction 51
    Gravitational Repulsion 56
    Propagation of Radiation 63
    The Construction of Matter 71
    Cosmic Expansion and Energy Conservation 77

Chapter 4. New Horizons in Cosmology 79
    The "Small Bang" Scenario 79
    Quasars as Celestial Beacons 83
    D/R Arcing of Radiation 94

Redshifts as Distance Indicators 99
  The Nearby Cosmic Edge 101
  The Enigma of Galactic Redshifts 106
  The Microwave Background 114

Chapter 5. Cosmic Reincarnation **118**
  The Great Cosmic Pyramid 118
  The Universal "Time Constant" 125
  Beyond Human Evolution 129
  A Cosmic Paradox 134
  Summary 135

Appendices **139**
  1. Inferred Separations of Galactic Associations 141
  2. Evolution of "**M**" Stages 142
  3. Extragalactic Hydrogen Clouds 143
  4. Gamma-Ray Bursts 145
  5. The "Missing Mass" Dilemma 146
  6. Calibration of Hubble's Constant 147

Part Two:
Religion and Society

Chapter 6. The Issue of Biblical Credibility **153**
  Inspired Prophets/Priests? 153
  Old Testament Authorship 155
  New Testament Authorship 157
  A Legacy of Unfounded Doctrine 159
  Biblical Symbolism 163

Chapter 7. The Old Testament **167**
  The Seven "Days" of Creation 167
  The Adam and Eve Story 169

Noah and His Ark 170
Some Early Biblical Narratives 172
The Saga of Moses 178
Examples of Supposed "Miracles" 183
The Book of Esther 187
The Trials of Job 189
Some Examples from Daniel 191
The Story of Jonah 193
Summary 193

Chapter 8. The New Testament **195**
The Enigma of Satan 195
The Jesus/Christ Paradox 196
The Genealogy and Birth of Jesus 198
Preparation for His Mission 200
Spiritual Significance of the "Miracles" 201
The Crucifixion and Resurrection 206
The "End of the World" Discourse 211
The Book of Revelation 214
Words of Lasting Wisdom 219

Chapter 9. Christianity and the Dead Sea Scrolls **220**
Discovery of the Scrolls 220
Dating of the Scrolls 224
Contents of the Scrolls 227
Christian Origins 231
Suppression of the Scrolls 235

Chapter 10. The Social Order **239**
Society and the Individual 240
Society and Religion 246
The Economic System 252
In Search of a World Society 260

Index **271**

List of Illustrations

Figures:

1. Divisions of Geological Time 5
2. Efficiency of Gravitation with Relative Motion 57
3. Distance/Repulsion Relationship of Gravitation 62
4. Redshift/Time/Velocity/Distance Scale *(Concentric View)* 68
5. Expansion of Universe and Propagation of Light 69
6. Observable Quasars in Terms of Z Redshift 82
7. Northern Hemisphere Quasars 86-87
8. Southern Hemisphere Quasars 88-89
9. Quasar Luminosity by Redshift/Direction 91
10. Redshift/Time/Velocity/Distance Scale *(Eccentric View)* 92
11. Galactic Arcing of Radiation 96
12. Pyramid Structure of Universe 133

Tables:

1. Average Redshift (per MLY) in Terms of Distance/Direction 108
2. Average "Center Group" Redshift (per MLY) 109

The Universe

Through distant ages man has gazed, with reverence, into the sky;

Foremost in his mind, that momentous question: Why?

How far are yonder points of light,

Shining like tiny beacons on the darkest night?

Whence came the stars, our heavenly host,

Their soft light shimmering like a phantom ghost?

Born of countless suns our Milky Way doth shine,

An enigma to ancient theories, which man must now refine.

Beyond the reaches of our Galaxy, in the enveloping depths of space,

Unnumbered starry systems participate in a truly cosmic race.

For the very universe is expanding, at a most prodigious rate;

Motivated, one would deduce, by Principle: A synonym for Fate.

– H.W.G.A.

Introduction

We find ourselves living in an age which is likely to prove unique in the history of our planet. Never before, in the brief span of a human lifetime, has so much knowledge been made available to those with inquisitive minds. Emerging from the depths of a dimly perceived evolutionary past, the guiding light of science shines as a promising beacon amidst the darkness of ignorance and superstition. But the influence of tradition often dies hard. An inherent reluctance exists to abandon an appealing philosophy or religious viewpoint, no matter how baseless or improbable. This has led to a number of disconcerting spectacles.

At the frontiers of science, where circumstantial evidence must frequently be used to test the validity of a premise, there is always the danger of jumping to false conclusions. However, unlike the case with religious dogmas, there is reason to expect that a scientific falsehood will eventually be uncovered and discarded. It is here, in in a willingness to admit past error, that lies the strength of the scientific method and which has facilitated the enormous progress of recent years.

The immense complexity of the universe has caused science to be divided into multiple fields of inquiry. This specialization is both a blessing and a handicap. On the one hand, it is essential in order to obtain deeper insight into a specific phenomenon. In contrast, it is also a valid point that some phenomena cannot be explained very well by a single arbitrary branch of science. In truth, a solution may encompass many seemingly diverse fields *or even have its roots buried in the realm of philosophy!* As a result, one must take care not to succumb to the "unable to see the forest for the trees" syndrome. There is a pressing need to mold a wealth of information into some comprehensive cosmic scheme.

Unfortunately, science has so far managed to present only a very fragmentary glimpse of a vast and mysterious universe. In doing so, it has managed to destroy the false religions and philosophies of many pious souls – but without providing a viable alternative! Such an ensuing vacuum cannot fail to leave many confused and disillusioned with regard to their faith in a meaningful creation. As events have turned out, there is presently little correlation of science with the idea of an underlying *Cause* or *Motivation*. This is a problem which we will endeavor to deal with in Part One of this book, where the popular but erroneous Big Bang cosmological model will be seen to be replaced by a truly dynamic cosmos - one featuring the principle of universal reincarnation and Continuous Creation.

A number of misconceptions, dealing with religion and the social order, are addressed in Part Two. In particular, it will be shown that the various biblical authors were far more concerned with the spiritual aspect of their stories than with actual fact, and therefore contain a hitherto unsuspected system of symbolism. Yet another deception will be seen to hinge upon the manner in which information about the Dead Sea Scrolls has been withheld by certain church leaders fearful of their revolutionary content. Finally, there is discussion of the social order, along with suggestions as to how many shortcomings may be rectified.

It is sincerely hoped that this work will assist one to construct a more realistic philosophy of life and to appreciate man's role in a dynamic and evolving universe. If so, it shall have accomplished a primary function. To the extent that it succeeds in stimulating a personal search for additional knowledge of such vital subjects, it will have fulfilled an even greater purpose.

– H.W.G.A.

PART ONE:
Science and Philosophy

Part One:
Science and Philosophy

1

Evolution and Reincarnation

The Procession of Life – The Evolution of Man – The Mechanism of Evolution – The Pyramid Concept of Reincarnation

Although modern theories of stellar evolution and cosmology are only able to define broad limits with regard to the age of the Earth, there is fortunately a way to obtain a date which is more specific. In a sense, certain rocks constituting our planet's crust may be said to contain a sort of internal clock, one remarkably trustworthy and capable of running for many billions of years. These rocks contain radioactive elements, such as U^{238} (92 protons, 146 neutrons) and U^{235} (92 protons, 143 neutrons). These two isotopes of uranium have half-lives of 4.51 billion years and 0.707 billion years, respectively. The former decays spontaneously to an end product of an isotope of lead known as Pb^{206} (82 protons, 124 neutrons); while the latter ends up as the lead isotope Pb^{207} (82 protons, 125 neutrons). Inasmuch as the natural isotope of lead is Pb^{204} (82 protons, 122 neutrons), appropriate analysis of rocks containing these uncommon isotopes is capable of yielding a fairly reliable guide as to the time that has elapsed since their solidification. Upon utilizing this procedure, scientists are now reasonably certain that the Earth's solid crust was likely formed about 4.5 billion years ago, which is in agreement with that of both meteorite and lunar studies and a parent Sun of slightly greater antiquity.

Conditions on our planet's surface must have been quite different in the very remote past. In particular, free oxygen is believed to have been a relatively scarce commodity, only obtaining its present abundance some billions of years later due to the build-up of oxygen-secreting plant populations. Lacking an oxygen-ozone atmospheric shield, with which to cut off much of the ultraviolet energy of the Sun, intense radiation would have bathed the surface of the Earth to a degree highly detrimental to advanced forms of life. By a remarkable coincidence, this is just what nature required in order to facilitate development of the earliest living structures. Even at this initial stage of evolution, one is caused to reflect whether simple chance happened to assist molecular chemistry – or whether it was really an intrinsic phase in some *Purposeful Cosmic Plan*.

The Procession of Life

As most people know, virtually all knowledge of ancient life stems from the finding of fossils. A fossil may be defined as the remains or impression of some plant or animal that lived during a past age. This information may be in the form of a shell or skeleton, a specimen in a state of petrification, an organism almost completely preserved in such a substance as amber, or merely an impression of a footprint imposed upon a layer of sandstone, etc. However, it should be borne in mind that only a limited number of living creatures can be expected to die in a manner conducive to fossilization and, due to a variety of factors, only the very smallest fraction of these will ever be found. Thus it will be seen that we can never hope to derive a full picture of all the myriad forms of life which once existed on the face of our globe. A reasonable idea as to the probable age of a fossil may be obtained by determining the age of the rocks in which it is found. (See Figure 1 for the divisions of geological time and a sample of the life characterized by each era.)

The earliest direct fossil evidence of life is thought to be some primitive types of microorganisms found embedded in rocks dating back as far as 3.5 billion years ago, suggesting that little time elapsed once the Earth had cooled to the point where delicate molecular bonds could prevail. In addition to microscopic bacteria-like organisms, blue-green algae of a complexity not too dissimilar to that of certain modern strains have been uncovered in rocks from the Lake Superior region of North America. Dates ranging from 2.0 to 2.7 billion years have been assigned to these finds, which indicates that cellular organisms had already achieved prominence at this distant age.

By the dawn of the Cambrian period, some 570 million years ago, a wide variety of marine life had become firmly established. Large numbers of trilobites, a sea-bottom inhabitant which bore a slight resemblance to a

Figure 1. **Division of Geological Time**

ERA	PERIOD & EPOCH		LIFE FORMS	YEARS (millions)
CENOZOIC	Quaternary	Recent	Primitive Man	2+
		Pleistocene		12
	Tertiary	Pliocene	Early Primates	26
		Miocene		38
		Oligocene	Mammals have replaced Dinosaurs	54
		Eocene		54
		Paleocene		65
MESOZOIC	Cretaceous		Flowering Plants	136
	Jurassic		Primitive Birds and Mammals	195
	Triassic		Dinosaurs	225
PALEOZOIC	Permian		Reptiles	280
	Carboniferous	Pennsylvanian	Insects	320
		Mississippian	Lush Forests	345
	Devonian		Amphibians	395
	Silurian		Land Ferns	430
	Ordovician		Many Fishes Trilobites	500
	Cambrian		Marine Life	570
PRECAMBRIAN			Invertebrates	700
			Simple Cells	
			Elementary Life Forms	3,500+

horseshoe crab, are found embedded in rocks of this particular time. Other common fossils include many types of mollusks, crustaceans and brachiopods. In fact, all major invertebrate phyla are fully represented at this stage of evolution. The advance to an internal bony skeleton is generally believed to have commenced at the beginning of the following Ordovician period, about 500 million years ago, as fossil remains of primitive vertebrates first begin to appear in rocks of this date. No trace of land life (plant or animal) is yet in evidence, as the sea continues to hold its exclusive franchise.

This state of affairs was to be rectified during the subsequent Silurian period, as certain forms finally commenced a transition from an aquatic to a terrestrial environment between 450 and 400 million years ago. Some elementary types of fishes also made an appearance at this time, and achieved great diversification and prominence by the close of the following Devonian period, about 350 million years from the present. While some species of early fish were able to develop external armor of solid bone, it was not until later that fish in general were able to turn their internal skeletons from cartilage into bone. It would seem that the more "progressive" amphibians, undergoing simultaneous evolution, were able to acquire bony vertebrae before the fishes themselves. In any event, the Silurian period was characterized by a number of subtle structural changes in emerging life forms – including features which would one day provide a basis for higher echelons of biological manifestation. Small fern-like plants and a variety of swampy growths were among the first vegetation to have stepped ashore, to be followed by an assortment of primitive insects, some of which resembled the centipedes and millipedes of today. The land was at last beginning to bring forth fruit.

The Devonian period witnessed the culmination of revolutionary changes that had begun millions of years earlier. Certain advanced forms of marine life managed to supplant their gills with an entirely new mechanism permitting them to take oxygen from the air: they had developed lungs. (It is instructive to note that several species of lungfish, which can breathe by means of either gills or lungs, remain even today as a reminder of a past age when life was emerging from the sea. Such forms also began to convert fins into leg-like appendages, endowing their owners with the ability to crawl.) The great climax to all these elaborate modifications, which had been taking place in the sanctuary of numerous shallow inland seas of this era, was eventually reached when the first amphibians invaded the land.

This higher life form possessed the ability to choose between a terrestrial or aquatic environment. In effect, the amphibians were now free to share the advantages and challenges of two distinct worlds – with this most recent acquisition literally opening the door to another universe: land!

Very noticeable changes occurred during the Carboniferous period, which began about 350 million years ago, and in the succeeding Permian period commencing roughly 70 million years later. Gigantic ferns, huge horsetails, club mosses and dense forests were typical of this age of lowlands and tropical swamps. Since much of this vegetation grew in regions of swampy marshlands, large amounts of decaying matter were soon covered with water and, robbed of the oxygen in the air, ultimately gave rise to deposits rich in hydrocarbons. Thus it is to the solar energy stored by the luxuriant vegetation of this distant age that we owe our deposits of coal and oil.

With the passage of time the amphibians were gradually superseded by a variety of reptilian creatures with superior size and speed. One additional advantage was advanced egg-laying capability. Unlike their predecessors, whose eggs required constant immersion in water, reptiles were able to produce eggs having a tough outer skin or shell which inhibited desiccation. Only at this point could it be said that the evolving animal phyla had truly severed the last remaining link with its aqueous origin. Appearing initially in the Carboniferous period, it was not until the following geological period that reptiles were able to achieve dominance. It is of note that the fossil record of this era also reveals some evidence of archaic forms of pre-mammalian life, suggesting an ancestry for mammals dating back to the earliest reptilian inhabitants.

Spanning an interval of approximately 100 million years (from about 230 to 130 million years ago), the Triassic and Jurassic periods account for some truly bizarre fossil specimens. Often referred to as the Age of Reptiles, the size and structure of some of these creatures are quite amazing. Unquestionably, the most publicized forms comprise the group of so-called dinosaurs, certain species of which grew into monstrosities as great as 100 feet or more in length and weighing over 70 tons! And yet, in spite of their enormous bulk, they possessed tiny brains out of all proportion to their size.

The honor of being the largest of these prehistoric monsters goes to a group of four-legged sauropods. *Brachiosaurus, Apatosaurus* and *Seismosaurus* are typical examples of a type of tree-browsing dinosaur with long necks and equally long and powerful tails. However, one of the most ferocious and fearful of these creatures appears to have been a species known as *Tyrannosaurus rex,* a giant kangaroo-like reptile over 40 feet long and standing almost 20 feet high. This monster was strictly carnivorous, as his powerful jaws and long, knife-like teeth testify. In fact, he could conceivably be referred to as an engine of destruction – a task for which he was certainly well fitted.

In self-defense against such enemies some reptiles developed armor plate, with the result that a number of species were virtually walking fortresses of sharp, bony plates and spines. One notable member of the horned and armored dinosaur category was *Triceratops,* a heavy creature built something like a rhinoceros, only much larger, often reaching 25 feet in length. The head of this animal was embedded in a huge bony shield from which two formidable horns resembling sharp spears projected. A third but somewhat smaller spike extended from the vicinity of its nose. Although vegetarian in nature, when aroused it must have posed a most effective battering ram.

The Jurassic period featured the appearance of a new form of life, a variety of flying reptiles which are generally classified under the name of pterodactyls. These winged reptiles had no feathers like modern birds; instead, their bat-like wings were formed of massive folds of thin, leathery skin. In size as well as in minor details these creatures varied widely; some were no larger than a bat while others had wingspreads equal to that of a small airplane! Judging by the structure of these larger pterodactyls, however, it is doubtful if their wing muscles were strong enough to allow them to fly in the manner of ordinary birds; they would seem to have been restricted mainly to gliding and soaring.

The suspicion that birds have a reptilian ancestry was confirmed with the finding of such fossils as *Archaeornis* and *Archaeopteryx* in rocks of Jurassic age. These creatures were a most peculiar blend of half reptile and half ordinary bird. Although their bodies seem to have been poorly covered, probably with little more than fringed or tufted scales, their wings and tail were unquestionably blessed with real feathers. Both of these early specimens were comparatively small, about the size of a common crow, and possessed reptilian heads equipped with rows of sharp teeth. But perhaps the most outstanding characteristic of all is to be found in the structure of their wings; for at the end of their wing bones projected several usable claws – rather striking evidence that their front legs had been converted into wings! One could hardly desire more suitable specimens to illustrate evolution from reptiles into birds.

During the subsequent Cretaceous period (about 130 to 65 million years ago) a number of dramatic changes occurred. In the plant world such modern trees as willows, poplars, birches and oaks began to displace the ancient cycads and conifers. The hot, tropical climate which had hitherto prevailed over much of our planet's land mass now began to moderate and the extensive swamps and lowlands underwent fluctuations and adjustments as the ground once more pushed upward in its most recent major upheaval. But the greatest surprise of all is to be found in the animal kingdom; for with the close of this period the seemingly invincible dinosaurs have literally vanished from the fossil record. Never in the course of geological history has there been such a swift and thorough extinction of established species.

This mysterious disappearance has been the source of considerable speculation, inspiring a variety of quite different theories. One proposal is that a nearby star exploded as a supernova, and that the ensuing radiation most adversely affected the larger terrestrial animal life forms. Another idea would seek to blame the extinction on an inability to cope with a changing environment. A more recent, and perhaps more likely, concept involves postulating an encounter with an asteroid-sized celestial body – from whence the Earth's atmosphere became so filled with pollutants that the vegetation upon which the dinosaurs depended was seriously depleted. Yet another possibility, as some authorities have suggested, involves the emergence of the smaller but more intelligent mammals. Was it, they hasten to ask, just a coincidence that the two events should have transpired at this same point in time? Could it be a case of intelligence prevailing over sheer brute force? (The reptilian practice of abandoning their eggs to hatch by themselves, for instance, could well have proven a disastrous trait in an age which featured smarter creatures with a liking for unguarded dinosaur eggs.)

In any event, with the conspicuous absence of the great reptiles the tiny mammals were now free to multiply and achieve dominion. The mammals at the beginning of the Tertiary period (about 65 million years ago) were typically not much larger than a cat or dog; yet despite their smaller size they represented an immense stride over the reptilian life that had preceded them. No longer are they vulnerable to relatively small changes in temperature and humidity, for their bodies are now covered with hair or fur and their blood is of constant temperature. In proportion to size, the structure of their brain is on a vastly higher level. Instead of laying eggs they bear their young alive. But what is even more significant, they remain with them; they nurse them and care for them until they are able to fend for themselves. A new era had dawned – a teachable form of life had come into existence, and with it had come the priceless advantage of learning by imitation and communication!

Over the course of millions of years the tiny mammals grew much larger and radiated into a wide variety of forms. Not all of these warm-blooded animals remained terrestrial, however, for just as a number of reptiles returned to the sea so did some species of mammal. If size is any indication of successful adaptation, then the whale would certainly seem to qualify, since it is by far the largest living creature. And yet, in spite of his fish-like form, there can be little doubt that the whale was once a terrestrial animal; for while it is now without legs a close investigation of his skeleton reveals vestiges of limbs (with five finger-like appendages) which have since degenerated to the point where they escape notice from the outside, but nevertheless are highly revealing from the inside. (It is of interest to note that vestiges of degenerated limbs may also be found in such drastically altered creatures as the python, a large snake.)

By good fortune, sufficient fossils have been found to afford a clear picture of the growth and development of a number of species over a considerable interval of time. Chief among the mammals are the horse and the elephant. The history of the horse can be traced back about 60 million years, when it was roughly the size of a dog and was a multi-toed animal suited for forest travel. Some 20 million years later it was a three-toed creature, with the middle toe larger than the other two. About 25 million years ago it had grown to the size of a small pony, and although it still retained three toes the two outer ones had so deteriorated that the weight of the body was borne almost entirely by the much larger center toe. With the passage of a further period of 12 million years, the fossil record reveals a virtually hoofed animal fully adapted to run on the many grassy plains of this era. (Incidentally, vestiges of these degenerated toes are still present today in the composition of a horse's foot.)

The earliest known ancestors of our modern elephant first appear in rocks of the Eocene epoch in Egypt, where it was then a small marsh or river-dwelling beast little more than two feet in height. During the Oligocene epoch it doubled its size and succeeded in acquiring both tusks and a primitive trunk before branching out into a number of subspecies. By the middle of the following Miocene epoch certain of these descendants – called mastodons – stood some five or six feet high at the shoulder and had managed to migrate, by means of the narrow causeway which connected Alaska to Siberia at this time, to such distant regions as the American plains. Pliocene rocks contain the bones of the larger mammoth, which was similar in many respects to our present elephants. Apparently the lower temperatures brought on by the Ice Ages served to encourage the growth of a warm covering of hair, as the species of elephant roaming large areas of North America and Northern Europe by Pliocene times were all huge woolly mammoths. (The frozen and almost perfectly preserved remains of this creature have been uncovered in Alaska and in Siberian tundras, where they perished sometime during the most recent Ice Age.)

The past history of the order of primates, which includes such animals as lemurs, gibbons, monkeys, apes, chimpanzees and finally man himself, is not so readily ascertained. Not being especially numerous they are not so likely to become fossilized. Nevertheless, a surprising amount has been learned. Fossil remains of tiny monkeys and lemur-like creatures have been found in rocks of Eocene date, as early as some 40 million years ago. Later rocks reveal increasingly larger and more varied species of primate, some of which bear a much closer resemblance to man than do the anthropoids living today. This fact clearly indicates that mankind did not descend from any existing species and strongly points to a common ancestry somewhere back near the dawn of the Tertiary period.

The Evolution of Man

Among the earliest primates, qualifying as an ancestor of modern man, could be the genus known as *Ramapithecus*. According to tests, involving the decay of radioactive potassium into argon, a specimen of this hominid found in East Africa has been dated back about 14 million years. The reduced size of his canine teeth, in particular, is deemed to argue for a reliance upon primitive tools and weapons. In turn, this would imply the utilization of hands and a necessity to stand and walk, at least at times, solely upon hind legs – making it possibly the oldest known biped among the primates.

A somewhat later and more human-like genus is *Australopithecus*. Fossil remains of this definite biped have been recovered from several sites, notably South Africa and East Africa, with the Omo Valley find in Ethiopia being conducive to dating by the potassium-argon method. This technique has yielded values ranging from almost 2 million years (in the upper strata) to over 4 million years (for fossils associated with lower strata). Not only is there strong evidence of tool usage, but there are even indications that he may have been a manufacturer of crude implements. It is one thing for a primate to pick up a digging stick or club; it is quite another to fashion items involving manual dexterity and intelligence. Since wooden artifacts are perishable by nature, it is not surprising that these early finds are monopolized by an assortment of chipped stone cutting tools – impressive evidence that the Stone Age had indeed dawned.

Zinjanthropus was discovered in the Olduvai Gorge region of Tanzania, East Africa, and has been dated at about 1.9 million years. The specimen in question possessed features less ape-like than older finds, having a deeper and more modernly arched palate and a cheek curved in almost the same shape as that of a human. Scattered about a related campsite, on the shores of what had once been an ancient lake, were a number of pebble tools of a form substantially more advanced than the earliest examples of Stone Age culture.

While the African continent seems to have held a monopoly with regard to oldest fossil evidence of primitive man, implying that it may have been a main focal point of human evolution, this picture is modified with the discovery of more recent specimens. *Pithecanthropus erectus* (Java Man), believed by some experts to have an antiquity of the order of 800,000 years, bears an even closer resemblance to man than older finds. *Sinanthropus pekinensis* (Peking Man), found in the depths of a Chinese cave, is thought to be nearly 400,000 years old. With a brain case not far below that of modern man, he was an accomplished tool maker and hunter. In fact, charred bone fragments and ashes indicate that, along with a more recently uncovered site in Hungary of about the same age, he was one of the first users of fire. Perhaps among the oldest of European finds, the *Heidelberg jaw* (found in 1907 in a sand and gravel quarry near the German village of Mauer) is considered to be a *Homo erectus* type roughly comparable in age with that of Peking Man. Yet another informative site, of still more recent vintage, is that of Terra Amata at Nice, France, where evidence of large communal huts has been uncovered. Constructed of wooden posts braced with stones, they each housed a small fire hearth. Not only has this site yielded a wealth of stone tools, but the impression of fur hides and even a human footprint was discernible to excavators.

Possibly an early form of *Homo sapiens* (present human species), *Swanscombe Man,* found in a gravel pit to the south of the Thames River in England, has been given a date of about 250,000 years. Judging by the assortment of associated tools and other artifacts, including flints and charcoal remains of campfires, it is logical to assume that he was a successful hunter who likely enjoyed home-cooked steaks. In many respects similar to the *Steinheim skull,* discovered in a gravel pit (of the same interglacial age) near Stuttgart in Germany, both specimens appear to be good candidates as ancestors to later European Neanderthal types. Perhaps an even more immediate relative of this next evolutionary plateau (with a probable antiquity between some 100,000 and 200,000 years) would be a group known as the *Solo skulls,* a slightly later type of early sapiens from a valley of the Solo River in Java.

Neanderthal Man tends to dominate the fossil record of primitive man from at least 75,000 years BP (before present) to as recently as 50,000 or even 40,000 years ago. Deriving their name from a find in a cave of the Neander Valley, Germany, this race is revealed as a rugged breed of hunters who once roamed over a substantial area of the world. Since the initial discovery a considerable number of similar remains have been unearthed, with some notable examples being *Rhodesian Man* (found at Broken Hill, Rhodesia), the *Hopefield* and *Saldanha skulls* (South Africa) and the *Ma-pa skull* (China). Adept in the art of fashioning a wide variety of cutting tools and weapons, their hunting prowess was sufficient to enable them to bring down animals ranging from deer to mammoth and rhinoceros. They also appear to have reached the point of entertaining religious feeling, since the arrangement of some of their bones clearly implies a ceremonial burial. In fact, in an excavated Swiss cave there was even evidence that one group tended to admire (or worship) stronger animal life forms – placing the skulls of bears, in this instance, in such a way as to invite interpretation as a shrine!

Once the age of a fossil falls below about 50,000 years, it then becomes eligible for dating by the carbon-14 isotope technique. Some rather specific dates have been established for a range of specimens utilizing this superior method. Intermediate forms between late Neanderthal and modern sapiens, his eventual successor, have been found in a number of locations and tend to reveal a continuum with regard to both physical evolution and cultural progress.

With an effective means of dating many otherwise uncertain finds, much information has become available concerning that interval just prior to the dawn of recorded history. In particular, the Dordogne area of France, with its beautiful valleys, limestone cliffs and caves, has provided much valuable insight. The Combe Capelle find, for example, has yielded dates of 34,000 to 25,000 years BP. The *Cro-Magnon* discovery (often depicted as the race which succeeded Neanderthal Man) is dated between 25,000 and 20,000 years BP. In other caves of this region a wealth of artifacts has been recovered, ranging from awls and needles (used in the manufacture of clothing) to talented artistic displays. In this last respect, literally hundreds of colorful paintings have been found on the cave walls. Chiefly depicting animals with which they were acquainted, they are of such quality that they would do credit to many present-day artists.

Migration of certain members of the Mongoloid race, from Asia to the Americas, is believed to have occurred during an occasion when Siberia was linked to Alaska by a land bridge. Moving southward, it was merely a question of time until they inhabited the far reaches of South America. Sites of human occupation of about 24,000 years BP have been discovered in the New World, and include such widespread locations as the Canadian Yukon, Mexico and Peru. In truth, the ancestry of mankind is revealed to be extensive in every sense of the word – utilizing virtually the entire globe as a nursery, *with his origin rooted at the very dawn of life!*

Invariably, the evolution of man clashes with a popular religious viewpoint. By tradition, Christian theologians have always vigorously opposed evolutionary theory in any shape or form, preferring to retain a blind faith in a literal interpretation of the Old Testament account of creation. Over the past century or so this official attitude has experienced little change, even if the more enlightened members of their flocks have long since entertained grave doubts as to the wisdom of ecclesiastical authorities. Some members of the clergy, unable to completely ignore overwhelming scientific evidence, would attempt to mitigate the situation by means of a compromise. In effect, they would begrudgingly accept the reality of evolution up to a point – namely, to the extent that man himself is not included in the overall scheme. This highly arbitrary exclusion is necessary, of course, in order that they might retain what must be described as a bizarre interpretation of the Adam and Eve narrative, along with an equally bizarre redemptionist doctrine requiring a mystical fall of man. (This aspect will be discussed at length in the second half of this book: *Part Two – Religion and Society.*)

In addition to conclusive scientific evidence, confirming man's evolutionary heritage, there is strong support from such sources as philosophy and common sense. From a philosophical point of view, the fossil record is one of ever-increasing complexity and intelligence. This is precisely the pattern needed to inspire contemplation of the dynamic principle of evolution – leading to the conviction that we must reside in a universe ruled by *Purpose*. Plain reasoning (common sense) is able to prove the evolution of man by the simple expedient of eliminating the sole alternative of "Special Creation." Other than by accepting the evolutionary premise of changing species, the staggering profusion of fossil entities can only be explained by postulating an *almost endless number of separate creations, extending over vast periods of time!* But if God be the Creator, why all the trials and tribulations and "experiments" before finally deciding upon creating the myriad forms of life existing today? In terms of "Special Creation" the whole affair is obviously preposterous. Evolution – *including that of man* – is capable of providing not only a far simpler explanation, but it is the only one which makes any sense at all!

The Mechanism of Evolution

How could such incredibly complex biological forms evolve from a microscopic origin? It is unquestionably a miraculous feat, and one which defies adequate scientific explanation, but if we possessed the faith of a "grain of mustard seed" the phenomenon of evolution would be seen to be a relatively minor achievement. For nature demonstrates daily the evolution of life into exceedingly complicated structures. From a tiny seed may spring a huge and towering tree; from a single living cell may evolve a creature comprised of many trillions of cells; where there was once a lowly caterpillar a beautiful butterfly emerges. All of this not only takes place before our very eyes, but it does so without necessitating one act of "Special Creation;" for essentially, *evolution is the mechanism used throughout!* Since it is obvious that these everyday "miracles" could not have occurred as a result of chance, we must admit the influence of "anti-chance" or God – an assumption which, incidentally, also explains the overall success of evolution in spite of numerous monstrosities and blind alleys.

As most people know, all living tissue (plant or animal) is built around numerous tiny structural units called cells. The average cell is quite small, usually only a few ten-thousandths of an inch across, and so cannot be viewed except under a microscope. But in spite of their limited size, cells are exceedingly complex entities containing many billions of atoms; they possess inherent properties common to all higher forms of life. They can eat, digest, secrete, reproduce and, in some instances, have even displayed an ability to learn through experience. They are alive by all standards of evaluation. Certain types of cells, such as amoeba, protozoa, infusorian and bacteria, are essentially single-celled animals living on their own; while others live together in colonies ranging from a few cells to structures comprised of vast multitudes. (An ordinary housefly may contain as many as several hundred million cells, while man himself is estimated to consist of hundreds of trillions of individual cells.)

The nucleus of a cell holds the secret of organization and control of the entity. Within this region are elongated strands of a nucleic acid molecule called DNA. Assuming the shape of threadlike spiral bodies known as *chromosomes,* at certain stages of cell division, these structures hold the heredity information essential for reproduction and for the overall functioning of the organism. In turn, chromosomes are comprised of numerous tiny segments of intelligence referred to as *genes* – each of which may be thought of as a miniature memory storage unit or "mind." Acting in conjunction with each other, these genes are able to produce a somewhat similar molecular substance named RNA, which is capable of moving out of the nucleus and acting as a form of "template" for the construction of the vital amino acids upon which life is based. On the whole, it may be said that the double helix pattern of DNA molecules – with their strings of genes – is highly analogous to a linkage of the vast swarms of neurons (brain cells) that constitute our own brains. Indeed, the only real difference is likely one of *magnitude* – with these minute genes interacting together to manifest an overall living entity just as surely as our cerebral structure (of linked neurons) is able to give rise to a human mind.

Thus it will be deduced that even the most elementary cell is a life form of incredible complexity, and one which could not have arisen from mere material particles were its evolution not guided by some intelligent force – *an intangible influence which must surely be ascribed to the presence of spirit!* Moreover, the dynamic nature of genes may be defined in terms of aggregations of atoms manifesting (or housing) a degree of intelligence , or spirit. And yet, the premise of specific concentrations of matter exhibiting spirit should not really come as a surprise; for after all, where there is life there must be spirit.

The means by which all living organisms have evolved into their present forms may be attributed to the effects of countless mutations. Insofar as evolution is concerned, a mutation can be defined as failure of the genes, inhabiting a germ cell, to reproduce themselves in an exact manner. No form of life, even comparatively simple organisms like viruses and bacteria, is free from mutations which appear to occur at random and are almost always harmful or useless variations. In general, changes of an advantageous nature are far too gradual to be observed over short periods of time.

But what, actually, are the changes that take place within the genes and chromosomes and which give rise to mutations; and what factor (or factors) may be held responsible for these changes? The structural changes which occur are of essentially two kinds. By far the most frequent cause of mutations must be attributed to losses, duplications or rearrangements of existing genes. However, from an evolutionary point of view, it is clear that constructive mutations are inevitably linked to internal changes within the constitution of the genes. Indeed, since no amount of duplication or juggling of existing genes could ever produce genes possessing different inherent properties, and endowed with the ability to perform more complicated tasks, it is obvious that the remarkable course of evolution rests primarily upon *internal* changes in the atomic configurations of the genes themselves.

When one surveys the path of evolution, with its great transition from the realm of atomic particles to incredibly complex structures, can it be seriously doubted that *guidance* was involved? Surely, the complexity of our very bodies is sufficient to convince us that we are not the product of chance. This is a conclusion which forces us to concede that some "intelligent" factor must reside within the genes and, moreover, that it is the actions and influence of this factor (spirit) which is largely responsible for determining the course of evolution – essentially, the nature of a constructive mutation. What might appear to be a matter of chance must, in reality, be ascribed to the actions of some hidden intelligence at work – *ultimately, to the influence of that Cosmic Principle we choose to call God!*

It is rather strange that a materialistic approach has dominated so many attempts to explain the mystery of evolution. What might best be referred to as the "mechanical" basis of most current evolutionary theories may be summed up under the terms of chance mutations, natural selection and adaptation. Upon such a flimsy platform purely random or accidental mutations are held to provide the raw material for the omnipotent and deciding influence of environment, which in turn is supposed to determine whether organisms possessing these new features will be allowed to live and to multiply. This process of weeding out undesirable and unfitted forms of life is known as natural selection or survival of the fittest. As a result the surviving forms are said to be those most suitably adapted to their environment.

Now this is all very fine, and we must certainly admit that an adverse environment will tend to eliminate malformed and unsuitably adapted species; but when we attempt to apply such arguments to the problem of evolution itself we find that, in actuality, *we have a framework capable of explaining essentially nothing!* Not only must it be acknowledged that advantageous mutations rarely (if ever) arise through pure chance, but the very course of evolution is plagued with an almost endless number of instances which are quite contrary to the dictates of natural selection. For example, it must be apparent that in practically every phase during the transition of life toward greater complexity, rudiments of useless or even detrimental physical forms must have arisen. One might say that some mysterious providence *foresaw,* as it were, the usefulness or *later need* of many diverse and highly complicated structures! Obviously, traditional "mechanical" theories are misleading and wholly inadequate, since they fail utterly to explain the course of evolution.

But the real problem for evolutionists may be traced to the first life itself. Where did it come from? Indeed, where does this "spark of intelligence" – representing the difference between life and death – continue to come from? For some unknown reason very few serious attempts have been made to incorporate the presence of *spirit* into a theory embracing evolution. Why this has been so constitutes a great mystery; *for does not all life – including our own soul – consist of spirit?* Furthermore, if we are to believe in the existence of a future life, and since there is overwhelming evidence as to the truth of evolution, we have no alternative but to recognize the presence and interaction of spirit with *all levels of creation,* from man down to the most elementary gene-complexes – even to the ultimate depths of matter itself!

A truly fundamental problem, which has long intrigued philosopher and scientist alike, is that of explaining a transition from inanimate molecular configurations to patterns characterized by life or spirit. When does a group of atoms cease to be inorganic matter and become a lowly form of life? Was the emergence of the first life on our planet simply a fluke of nature, involving a chance "fortuitous concourse of atoms?" Or, on the other hand, was it an inevitable result of dynamic properties inherent in the very constitution of so-called material particles?

Laboratory experiments have left no doubt as to the ability of atoms to engage in a spontaneous generation of organic molecules. Placing various common elements and simple inorganic substances in a flask containing water and autoclaving the entire apparatus so as to ensure sterility, scientists have performed tests utilizing such energy sources as heat, ultraviolet and X-ray radiation, electric spark, etc. The results turned out to be both impressive and quite conclusive, as a host of organic compounds were readily synthesized. In short, a wealth of complex organic chemical compounds is bound to be the rightful heritage of all newly-formed Earth-like planets. The widespread evolution of biological life must, accordingly, be deemed a foregone conclusion.

Conceding the inherent potential of matter to initiate life, the question remains as to how particular arrangements of atoms could give rise to intelligence. Not only do we have the level of microscopic gene-spirits and cellular entities to consider, but there is also the overall consciousness of the entire animal phyla – of which the human soul is only a solitary example – demanding similar clarification. Obviously, the whole issue hinges upon a structural positioning of atomic particles so as to form specific patterns conducive to the manifestation of spirit. Bearing in mind the non-spatial nature of spirit, it is perhaps useful to visualize this association with matter in terms of a "capturing" or "housing" hypothesis. That is to say, when certain intricate "fields of influence" have come into being, as a result of the formation of appropriate molecular configurations, these complex fields could be thought of as attracting and capturing a level of spirit or life. Thus the intelligence of gene-spirits would be seen to arise from specific atomic/molecular patterns; while the overall personality of a unicellular organism may be interpreted as being due to the larger scale patterning of gene-type configurations. Upon a still higher plane, as we contemplate more advanced animal life forms, the interactions of specific patterns of brain cells are able to facilitate the capturing of spirit of steadily increasing value.

In spite of a wild profusion of biological forms, over the lengthy course of evolution, there is clear evidence of *guidance* ever urging an upward development of life. This is evident at every major stage of diversification. In addition to explaining the construction of a multitude of exceedingly complex internal organs, which are totally inexplicable on the grounds of pure chance, this premise is borne out by a remarkable similarity in the external appearance of evolved life forms. For example, one has only to look at the broad spectrum of the world's mammal population. Is it just a strange coincidence that virtually all such creatures possess one head, one mouth, one nose, two eyes, two ears, four appendages in the form of pairs of arms and/or legs, five fingers and five toes (or claws), and so forth? Much rather, this parallelism can be deduced as a striving toward some *ultimate purpose* – namely, *Homo sapiens, and beyond.*

The highly diversified and somewhat haphazard path described by evolution can only be resolved upon the basis of free will. For it is solely by ascribing freedom of will to spirit that we are able to account logically for the numerous trials and errors of evolution. Failure to recognize free will as an intrinsic attribute of spirit is really tantamount to accusing God of outright incompetence in the creation of life! Imperfection exists in the universe simply because the many levels of spirit are less than Perfect – *not as a result of any direct action on the part of a Creator!*

Adopting this line of reasoning, the outcome of evolution may be acknowledged to hinge upon *probability.* That is to say, a specific proportion of spirits can be expected to utilize their potentialities and freedom of will in such a manner as to follow specified courses. Just as small-scale uncertainty of individual atomic particles gives way to large-scale stability when great numbers are considered, so nature has contrived to ensure overwhelming probability that a minute (almost infinitesimal) fraction of spirit would pursue paths that would terminate in man; while the vast majority could be relied upon to diverge and lead to the creation of an extensive chain of lesser species. Subsequently, our planet's biological life pattern *may be likened unto the structure of an enormous pyramid* – extending from myriad swarms of gene-type spirits to the present apex of modern man with his relatively small numbers.

The hypothesis of *Divine Guidance,* as the underlying motivation behind the success of evolution, admittedly places the entire problem of mutations and adaptation into the realm of philosophical deduction. But this in no way negates the validity of a concept. History is, in fact, permeated with instances in which philosophy and logic have preceded a scientific explanation. Contemporary science, with its roots still embedded in archaic "materialism," is just in its infancy when it comes to comprehending *Cause* or *Principle.* For this reason it is sufficient to theorize the existence of a subconscious "feedback" mechanism between the gene-complexes of germ cells and the overall controlling spiritual entity. While the traditional view of natural selection may be expected to play a role in the adaptation of any species to its environment, it is also clear that mutations are – in some vital sense – most certainly *directed!* Comprised as our bodies are of several distinct levels of life or spiritual awareness, it is not surprising that one manifestation should have no direct conscious association with another. Indeed, it is as though quite different worlds are involved, with each being able to indirectly affect the other by means of diverting guidance or virtue received from above – the ultimate source of which must originate in God.

Equivalent in many respects to that of conscience in man, the use of free will by the gene-spirits of germ cells must be held largely responsible for evolutionary mutations – producing both monstrosities (e.g., the dinosaurs) and the much more advanced life form epitomized by *Homo sapiens.* This same principle also applies to the ability of species to undergo environmental adaptation. What probably transpires is that *need* for structural change is capable of impressing itself upon the gene-complexes of germ cells. Exerting a "pressure" for a certain mutation (if at all possible), the Darwinian postulate of natural selection is then permitted to act upon favorable variations. (Examples of this adaptive capability in man range from the increased resistance of some races to specific diseases, to differences of skin pigmentation in others commensurate with their exposure to sunlight.)

Notwithstanding our present inability to grasp many details of biological evolution, it is both desirable and expedient to formulate one basic conclusion: *We must presume that the fundamental nature of the cosmos is such that all life contains spirit, and that it is able to rise to higher levels!* Only by making this profound assumption can we reconcile the idea of a future life with the reality of evolution. To think that man alone is the only spirit privileged to evolve to a higher status is absurd, since it is hopelessly in conflict with our knowledge that even man himself originated from a lower form of life – from a form far lower than the simplest cell, in fact!

The Pyramid Concept of Reincarnation

A useful definition of the term "spirit" is perhaps in order at this point. Stripped of associated attributes of intelligence and memory, it may well be constructive to define the broad spectrum of spirit as *specific and varying segments of Perfection,* or whatever God consists of. In essence, a spiritual entity is basically a particular *degree of unselfishness* – nothing more! It is surely this feeling of *Universal Love* – expressed as a "level of reality or righteousness" – which determines the actual status that a spirit is given, in the sense that the higher this level the more advanced its status with regard to the evolutionary structure of the universe. Thus it will be seen that spirit cannot be divided into classes or categories. On the contrary, during the long course of evolution our planet must have witnessed a steady and unbroken procession of spirit of gradually increasing worthiness or degree of Perfection. *Principle,* in which spirit forever seeks to equate intrinsic worth with physical status, is acknowledged to be the one dominant cosmic feature – completely transcending such mundane characteristics as space and time should this be necessary to preserve Harmony and Justice. Once again, it will be emphasized that – due to evolution – *no one spirit is able to progress to a higher state of manifestation to the exclusion of all others!*

Such assumptions raise rather pertinent questions which have been asked by philosophers over the ages. In brief, why is it that we find ourselves born into an imperfect world? Why, in fact, should any form of life find itself situated where it does? If the universe is the handiwork of an Omnipotent and Perfect Creator, why is anything at all imperfect? What, in the last analysis, is the deeper significance of ascribing free will to spirit? Invariably, the issue must hinge upon *justification of status* – in which spirit is seen to be not so much created as it is *constructed* in a lengthy step-by-step process from some infinitesimal origin. In order to acquire a higher nature, all spiritual entities must *earn or deserve* this most cherished reward. Interrupted upon innumerable occasions by death, the evolution of a spirit (or soul) may only proceed on the basis of what might well be described as the principle of *Cosmic Reincarnation.*

Actually, the advantages inherent in the concept of reincarnation are extremely convincing. The old biblical notion that we have but one all-determining life in which to prove ourselves worthy of passing to "heaven" is seen to be quite illogical. For example, what happens to the soul of one who dies in infancy? Should we be so naive as to think that it passes automatically to a higher existence we are faced with the thought that those who live longer are unlucky, inasmuch as they have risked not making it, so to speak! Furthermore, at what specific instant of time does a child cease being a child and become responsible for its actions? Clearly, no time limit whatsoever can be established. A similar parallel exists with regard to adults. Who is to say that time will not lead to profound changes in one's spiritual outlook? And yet, we know that death shows respect for neither time nor person. Are we to relegate Divine Justice to the whims of blind chance, by failing to give all individuals an equal length of time and equal opportunity to evolve spiritually?

Upon the premise of reincarnation such objections will be found to have a simple and common solution. By acknowledging the immortality of soul, or spirit, the time limit impasse is at once removed. The mystery surrounding premature death (or any death, for that matter) no longer exists, since an appropriate rebirth will prevent the injustice which would otherwise ensue. To an increasing number of thinkers, a belief in reincarnation offers the only alternative to the spectacle of a chaotic universe. But what of the laws regulating this most dynamic principle? Can this concept be related to evidence revealed by science?

Coexistent with the idea of reincarnation is the highly relevant phenomenon of evolution. Although feared initially by religious leaders in general (and still feared by many) as a materialistic doctrine, evolution was soon accepted by the scientific world as an undeniable truth. However, a major underlying reason for this religious rejection continues to invite more detailed clarification. On a philosophical basis, how are we to reconcile immortality of the human soul with certain inescapable implications of evolution? If man really did evolve, from at least a microscopic origin, then where are we to draw the line as to the specific stage in which he suddenly became blessed with this great potential? Most certainly, precisely the same problem arises as must accompany any notion of one all-determining existence. A little reasoning is sufficient to show that the two are indeed connected, *and that the answer to the one difficulty is also the solution to the other!*

Rather than strive to impose a time limit where none may be imposed or draw a line where none can be drawn, we are led to only one sensible conclusion. Without a doubt, it simply boils down to a case of *all or nothing* with regard to the issue of a life after death. Either all forms of life are eligible for a future rebirth, or else it is a false hope that does not exist at any level – including that of man himself.

Should we embrace the concept of a Just and Purposeful universe – in which all levels of spirit are reborn – the question of status becomes of paramount importance. For in order to justify its actual position upon the evolutionary scale of life, a spirit must *deserve* the level which it is given. Accordingly, it can only deserve a particular status if it has had a *previous* existence! This leads, inescapably, to the conviction that all spirit must be traced to an extremely low and common origin – from whence the reincarnation process has since achieved wide diversification from the elementary gene-spirit entities of initial life forms.

An extension of this assumption to the depths of matter is not without similar advantages in resolving the paradox of a Perfect Creator and an imperfect physical universe. As a matter of fact, considering that there is often great difficulty in distinguishing between animate and inanimate forms toward the lower extremities of life, any distinction is frequently more arbitrary than not. It is, in reality, almost tantamount to attempting to draw another line where no line may be drawn. It may well be prudent, therefore, to consider even lowly matter to be of the essential nature of spirit – with the only genuine difference between life and seemingly inert matter being *one of degree or value.*

Upon reflection, the concept of reincarnation appears to be in full agreement with a spiritualistic interpretation of matter, in spite of numerous instances of atoms and material particles being torn apart and fused together by circumstances over which they have no control. Invariably, the formation of new physical bodies must induce appropriate changes in associated spirit entities, in order that status (or degree of evolutionary progress) may again be justified. The process, in this case, is much the same as the phenomenon of birth and death in our own world. About the only difference which may be cited is that we are actually able to detect the birth of new entities in matter; whereas with the death of biological life we can see no direct proof of continued existence. It is surely a stroke of irony when the indestructibility of lowly matter may be adduced in support of the principle of spiritual immortality!

Proceeding with the fundamental assumption that spirit is capable of rising to immense heights, from the depths of obscurity, we are at once confronted with an intriguing observation. It so happens that the further one goes down the scale of life the greater the numbers encountered of the lower spiritual entities. The difference in the abundance of the many species of biological life alone is positively staggering – to say nothing of the vast multitudes of atoms comprising the simplest cell. *How could they all hope to evolve, individually, to higher levels?* For if these countless swarms of lesser spirits are blessed with the potential to eventually progress to the level of man (at least) – and they must be, since man himself is the product of a less-than-microscopic beginning – then does it not appear that only a fantastically small proportion could ever reach this advanced status?

There can, of course, be only one sensible conclusion: *We must deduce that many spirits of a low order may – in conformity to some cosmic law – be combined to produce fewer entities, of a higher overall nature or status!* Solely by recognizing the principle of *fusion,* as a vital and dynamic feature of the universe, may we escape an otherwise absurd situation.

The mechanism of this inferred fusion process urgently invites detailed clarification. Meanwhile, there seems no compelling reason why we could not theorize a useful picture in harmony with logic and observational evidence. (Inexplicably, the deep significance of this *fusion* principle has been overlooked by scientists and philosophers alike, with the result that a most important clue was missed – one which could have facilitated a much earlier solution to certain of nature's more treasured secrets.)

Let us begin by designating the lowest possible segment of spirit (or quantum of energy) as an "**A**." The least degree of fusion that it may enter into is union with another entity of like value. Should the two sufficiently deserve the resulting status they will surrender original identities in order to give birth to a new and higher overall entity (to which we might assign the symbol "**B**"), of twice previous individual worths. Similarly, the fusion of two "**B**s" could be viewed as leading to the formation of one "**C**," etc. The spiritual cosmos, upon the basis of such a doubling concept, might be pictured as a series of "steps" extending to higher levels of existence, *with each successive "step" containing an equal value of "A" but only half as many entities.* (So effective is this doubling process that a mere 100 "steps" is roughly equivalent to the huge sum of 10^{30}.) This is not to imply that the incidence of fusion is restricted to twofold jumps in status. On the contrary, there is every reason to presume a wide variety of intermediate complexes. For instance, the union of an "**A**" with a "**B**" must lead to the creation of an entity with a value of three "**A**s." The fusion of a "**C**" with a "**B**" would produce a structure valued at six "**A**s," etc.

Very frequently, in the oftentimes violent world of physical events, circumstances will produce an entity of a status undeserved by its recently-fused components. In such an instance it must bring about the "death" – and subsequent rebirth – of the separate entities on the grounds that they are unworthy of the new status so created. Spirit of the appropriate level, itself the product of a previous disruption, would immediately inhabit the physical structure of higher value. (Time and space simply do not have any meaning to spirit that is temporarily lacking manifestation.) In this manner the ends of justice will always be served within the framework of a Cosmic Scheme allowing a reasonable predetermined margin of tolerance.

By way of analogy our mysterious universe might be likened unto a vast pyramid, with rows of "steps" leading to its upper reaches. The base of the structure is wide, representing numerous entities of low worth, tapering steadily with height toward a point signifying fewer and fewer spirits of higher and higher value. At the top of this *Great Cosmic Pyramid* we have a peak or pinnacle which might be held to depict a state of *Maximum Fusion, or Oneness with God.*

Acknowledging that the basic principles of evolution and spiritual fusion must combine to produce a steady upward flow of spirit from lower levels, we may well ask where this dynamic process is likely to terminate. Regardless of the momentous and startling conclusion which must arise, no valid excuse can be advanced to predict an end to the evolutionary scheme only part way up the Cosmic Pyramid. This being so, we are left with no alternative but to recognize the inherent right of all creation to rise to the height of Perfection, or Total Fusion. For it must surely follow that, ultimately, *all spirit – man included – is destined to terminate in none other than God Himself!* Truth, or fusion into the One Body of God, is invariably our glorious Destiny! Therein is to be found both the Purpose and the Motivation behind creation.

2

Our Physical Universe

The Enigma of Matter – Solar Systems – The Stars – Degenerate Stars – The Milky Way – Galaxies of Stars – The Expanding Universe – Quasars – The Creation Impasse

Knowledge of the physical structure of our universe is being pursued by science upon two basic fronts – microcosm and macrocosm. On the frontier of microcosm, physicists have succeeded in tapping – for better or for worse – the enormous energy of the atom. On a purely beneficial note, they have also managed to unravel the mystery of the stars, providing valuable insight into the energy processes transpiring deep within their interiors. Concerned with the world of macrocosm, astronomers and astrophysicists have thus been greatly assisted by research in a branch of science endeavoring to probe in precisely the opposite direction.

Such an instance is typical of the interrelationship which exists among many seemingly diverse fields of inquiry, and affords a graphic illustration of the folly of those who would attempt to exclude all philosophical deduction from the realm of science. As long as this approach is capable of explaining phenomena, and is conducive to periodic testing of theory with observation, it provides just as sound a basis as the circumstantial evidence accepted by science in general.

However, before proceeding with this viable principle, which would seek to uncover the *Motivation* or *Purpose* behind phenomenon, it is necessary to examine certain aspects of our physical universe. Only through an understanding of the laws regulating so-called material structures – ranging from infinitesimal atomic constituents to giant galaxies of stars – is it possible to construct a workable hypothesis that would recognize spiritual motivation as being associated with lowly matter.

The Enigma of Matter

Throughout the ages, matter has commonly been looked upon as an inert and absolutely dead substance, and certainly as having nothing whatever to do with the spiritual aspect of the cosmos from which it was considered to be quite divorced. While, for practical purposes, this contrast in the nature of matter and spirit may well have some justification, scientific advances are tending more and more to raise concepts of matter to something above previous rock-bottom estimates – to the status of an enigma, if nothing more. For as strange as it may at first seem, the deeper science probes into the secrets of matter the less "materialistic" it actually appears to be.

Take, for example, a tiny molecule of oxygen in the atmosphere we breathe. Who would dream for one moment that this minuscule object, measuring only about a hundred-millionth of an inch, is constantly darting around and rebounding off other molecules with a velocity of something like a third-of-a-mile per second? Although many other atoms are not nearly so lively in their habits, they nevertheless are all in a constant state of agitation, ranging from high velocities for gaseous substances down to a more or less rapid vibrating motion for the most rigid solids. Who can honestly consider matter inert and "dead?"

From a dimensional point of view, the structure of an atom may be said to consist primarily of empty space! That is to say, its center or nucleus (manifesting positive charges of an influential force known as electricity) is about 10,000 times smaller than the diameter of the atom itself which is extended by a specific number (depending upon the element in question) of much lighter concentrations of mass termed electrons. Bearing negative charges of electricity they revolve, so to speak, around the nucleus. Although roughly analogous, in a sense, to that of incredibly minute solar systems, a major difference lies in the realization that electron orbits are constantly shifting so as to envelop the central nucleus in a series of concentric shells. Further evidence has resulted in the hypothesis of electron spin, in which these tiny particles are seen to resemble rapidly spinning tops – in addition to behaving like miniature magnets by virtue of possessing a negative electric charge. Typically, an atom consists of a central nucleus, about 10^{-12} cm in diameter, surrounded by an electron shield occupying a sphere which is roughly 10^{-8} cm in diameter.

The chemical properties of the various elements can be explained on the basis of the number of electrons in the outermost shell. For instance, an atom requiring one more electron to complete its outer shell will readily form a strong bond with another having a solitary electron in its most distant shell. (Thus sodium chloride, commonly known as table salt, is formed when a sodium atom with a "surplus" outer electron is united with a chlorine atom which needs a single electron to complete its outer shell. Two otherwise "aggressive" elements are, in effect, enabled to neutralize each other by reason of their ability to share an outer electron.) In turn, the number of electrons (with negative charge) characterizing a given atom will depend upon the quantity of protons (with positive charge) in the nucleus, since it is a cardinal rule of nature that unlike electrical charges attract while like charges repel.

Hydrogen, the simplest element, has a nucleus of but one proton and one encircling electron. Helium, the next heaviest element, has two protons and a completed single shell of two electrons, which accounts for its inert, or neutral, chemical properties. Uranium, at the other end of the scale of natural elements, possesses 92 protons and 92 encircling electrons. The clue as to how numbers of positively charged protons could reside together in a nucleus – without their like charges causing them to fly apart – was provided by the discovery of the neutron. Still constituting somewhat of a mystery, it is fully evident that they must act as a bonding medium to hold a compound nucleus together. (By way of illustration, a helium nucleus consists of two protons and two neutrons – with the neutrons serving to negate the mutual repulsive force of two positively-charged protons, while yet permitting them to attract two relatively distant and oppositely-charged electrons.)

Essentially stable when bound in atomic nuclei, a free neutron is radioactive and will transform spontaneously into a proton and an electron (plus a small amount of infinitesimal quanta, often referred to as neutrinos) after a half-life of about 15 minutes. Electrically neutral, the neutron has a mass equal to roughly 1,839.0 electrons, being slightly greater than the proton which is rated at about 1,836.5 electron masses. This ability of a free neutron to eject an electron – along with lesser bundles of energy – must surely be of significance to an understanding of the basic structure of matter. It does, in fact, provide ample proof that material particles are composed of varying amounts of "congealed energy" that has simply become fused together. Most assuredly, matter is of a divisible nature.

Instead of our being able to resolve matter into a few inert and unchangeable substances, as was once supposed, we now find that our physical universe is built out of what might best be described as divisible and interchangeable condensations of energy or forces – of *"fields of influence"* of varying strength. Clearly, it is a medium which defies explanation upon a purely mechanical basis and which has properties quite in contrast to the "dead" substance that our most imperfect senses would have us believe. It has often been said that electricity is a basic commodity out of which our universe is built. But just what is electricity? Moreover, what is the equally enigmatic force of gravitation? Perhaps the conspicuous failure of science to resolve such mysteries is an indication that an entirely new approach is in order – presumably, one in which *Principle* or *Motivation* is seen to play a prominent role. After all, at the level of biological life, does not mere physical matter succeed in manifesting what might be termed *spirit* or *soul?* Has science been guilty of inadvertently drawing a line between matter and spirit – when, in truth, *no line can be drawn?*

To summarize, it may be stated that the two fundamental particles of matter, which alone possess the ability to exist in a free and stable state, are the proton and electron – or their oppositely charged counterparts, the antiproton and positron. Extremely rare in our region of the universe, these counterparts have only a very transient existence, being annihilated in collisions with common matter. Positrons, for example, will readily unite with far more numerous free electrons to produce binary pairs of high energy gamma rays. For some very good reason, which must be ascribed to the actual circumstances surrounding creation, virtually no antiprotons were allowed to survive. Likewise, relatively few free positrons have been permitted to inhabit our region of space. This is just as well, of course, since the presence of large numbers of these entities would lead to a chaotic annihilation of material particles and a universe consisting of little else than pure energy. Invariably, this biased production and distribution of fundamental particles must be given due consideration by any theory attempting to account for the origin of our physical universe.

Solar Systems

As we all know, the Sun occupies the central position in our Solar System. Essentially a huge ball of intensely hot gas, it has a surface temperature of almost 6,000 °C and an internal temperature of close to

Our Physical Universe

13,000,000 °C at its center. The Earth's own diameter of almost 8,000 miles appears rather paltry in comparison to that of the Sun's 864,000 miles. In mass or weight, it is roughly a third-of-a-million times greater than the Earth, containing (by weight) about 75% hydrogen, 20% helium and approximately 5% of all the remaining heavier elements. The density of the Sun varies considerably at different depths, in the sense that the central regions are exceedingly dense while the outer layers are comprised of gases existing in a somewhat rarefied state. Although the Sun is some 93 million miles distant from us, it is not its particularly high luminosity but rather its comparative nearness which is responsible for its appearing so much more brilliant than the stars. Indeed, as great a contrast as there may seem to be, our Sun is simply *an average star!* It just happens to be the closest.

Of all the planets in our Solar System, only the Earth is suitable for the evolution of higher life forms. Mercury, the closest body to the Sun at a scant 36 million miles, is practically a facsimile of our barren and airless Moon. Venus, almost the size of the Earth and the next planet in order from the Sun at a distance of some 67 million miles, is characterized by extremely high surface temperatures and a thick carbon dioxide atmosphere many times denser than our own. Mars, the last of the inner planets at some 1-1/2 times the Earth's distance from the Sun, is about 4,200 miles in diameter and lacks sufficient gravity to retain much of an atmosphere. There is little hope of life having ever developed much beyond a microscopic stage. Jupiter and Saturn, the gas giants of our system with masses some 317 and 95 times that of Earth, orbit the Sun at mean distances of 482 and 888 million miles, respectively. Uranus, almost 15 times as heavy as the Earth, is a little over 19 times farther from the Sun. Neptune, the last of the semi-giants of our planetary family at a distance of close to 2.8 billion miles, has a weight slightly more than 17 times that of our Earth. At the very fringe of the Solar System lies the planet Pluto, in an eccentric orbit which carries it as far as 4.6 billion miles from the Sun. Now known to be smaller than the Moon, it must have a surface temperature that is perpetually hundreds of degrees below zero.

There are two other phenomena, belonging to our Solar System, which deserve special mention on the grounds that they contain clues to its origin. Lying chiefly between the orbits of Jupiter and Mars are swarms of smaller bodies known as asteroids. They range in size from giant Ceres, some 480 miles in diameter, to the dimensions of mere boulders and pebbles. Such bodies represent a combination of primordial material which simply failed to form another planet and evolving bodies that collided and disintegrated during the early history of our Solar System. The second phenomenon is that of comets, many billions of which are believed to circle the Sun as a distant halo some tens of billions of miles out in space. Comprised of frozen gases and bodies of solid matter, they constitute remnants of the epic process of planetary formation that took place shortly after the Sun's birth some 5 billion years ago.

Since one of the original discoveries about our Solar System was the knowledge that all of the planets revolve in the same direction as the Sun rotates, and on virtually the same plane as the Sun's equator, it was inevitable that this remarkable coincidence should inspire an idea as to the origin of our family of planets. Accordingly, it was not long before the theory was advanced to the effect that the Sun and planets were formed, at about the same time, from the condensation of a huge cloud of interstellar gas and dust. In conformity to a firmly established principle of physics, whereby contraction of a rotating body must lead to an increase in its rate of spin, it was assumed that when our protosun had shrunk to a critical degree strong rotary forces must have caused periodic ejections of material – from whence the planets and their moons later condensed.

Unfortunately, as is sometimes the way in science, the obvious solution begins to run into trouble when additional (but only partial) information becomes available. In this case, it was eventually shown that the Sun, containing more than 99% of all the matter in our Solar System, actually possesses only about 2% of the total angular momentum – the property that keeps the Sun rotating and the planets revolving around it. How was it possible, critics wanted to know, for planets having less than 1% of a system's mass to acquire a staggering 98% of its angular momentum? In effect, to balance the mathematics of the system it would be necessary to either increase the Sun's rotation by an enormous factor, or else have the planets orbiting at a very small fraction of their present distance. So impressive is this objection that, for some decades, it was deemed sufficient to completely rule out the idea that planets could have formed from a rapidly rotating protosun. Consequently, a variety of alternate theories were quite seriously entertained for over half a century, all of which relegated the formation of planetary systems to the status of a cosmic fluke.

But it was to prove a classic example of a little knowledge only serving to obscure an issue. The clue to an understanding was finally uncovered through a study of stellar rotations, when it became evident that some external force must have acted to drastically slow the Sun's rate of spin. From a study of recently-formed stars it soon became clear that rotation periods of a hundred times or more were the rule rather than the exception. What had caused the Sun's rotation to be reduced from hours to almost a month? It is now generally accepted that a combination of radiation pressure and magnetic fields act to push away rotating rings of primordial gaseous material from the vicinity of young stars. Thus it would appear that the condensing planets of our Solar System were repelled to considerable distances – at the direct expense of the Sun's loss of rotary momentum and a sharp reduction of its once-powerful magnetic field.

As a consequence of this discovery, it must surely follow that vast numbers of other planetary systems are bound to have arisen in a similar manner. Even allowing for the likelihood that many of these potentialities will be wasted, for one reason or another, it is quite impossible to escape the conviction that at least a good proportion of them will result in the formation of planets comparable to our Earth. The total number of probable abodes of biological life must almost certainly be in the billions within the confines of our own Milky Way alone, and billions of billions within the enormous expanse of the entire universe! As to the question of whether life similar to our own will develop on these other worlds, we have only to ponder the notions of certain misguided ancestors who once thought in terms of a flat Earth situated at the center of the universe. Clearly, this inescapable conclusion must dominate all future philosophical inquiry into the nature of the cosmos.

The Stars

Moving away from the limiting dimensions of our Solar System, we now turn to the fascinating subject of the stars. The method used in determining the distances of the nearer stars is really simple in principle. As a result of the Earth's revolution around the Sun, nearby stars will appear to describe a slight backwards and forwards motion with respect to more distant background stars. By means of utilizing this annual displacement (or "parallax"), it can be shown that even the closest star is quite remote – some 25 million million miles away. In order to more conveniently express such distances, the term "light-year" has been adopted. A light-year depicts the distance covered in one year by a ray of light traveling at the speed of some 186,000 miles per second, or almost 6 million million miles. In this manner the nearest star has been found to be 4.3 light-years away while Sirius, the brightest star in the sky, is at a distance of about 8.6 light-years.

Once the distances of stars have been measured, it then becomes feasible to determine their true brightness or "absolute magnitude." When this has been done such comparatively distant stars as Canopus, Rigel, Deneb and Betelgeuse are actually found to be many thousands of times more luminous than our Sun; while on the other hand many of the nearer stars are much fainter. (Proxima Centauri, at a distance of little more than four light-years, is only one ten-thousandth as bright as the Sun.) Thus it will be seen that the apparent or visual brightness of a star is no positive guide as to its true distance, since the stars themselves differ enormously in their luminosity.

The range in the sizes of the stars is exceedingly large. While most stars do not differ too greatly in diameter from our Sun and belong to what is known as the "main-sequence" (e.g., stars that have undergone little evolutionary change), there are some so large that if one of them were to replace the Sun, the orbits of the inner planets would actually lie within the body of the star itself! Although in general these huge stars have relatively low surface temperatures, their enormous sizes are sufficient to make them objects of very high luminosity. Such stars may exceed the Sun's diameter by literally hundreds of times and are known as "red giants" and "supergiants." At the other extreme are stars so small that while they possess the mass of stars they resemble the planet Earth in size. Objects falling into this category are referred to as "white dwarfs." As if this is not enough, even more compacted structures have been found to exist in the form of "neutron stars," typically about 10 miles in diameter and twice the mass of the Sun!

Despite the wide range in the sizes of the stars, their masses show surprisingly little variation, with the vast majority having masses between one-fifth and five times that of the Sun. Hence, it follows that the stars must display a most remarkable variety of densities. While red giants may have an average density that is less than our own atmosphere, the much smaller white dwarfs are so dense that a cubic inch of their material must weigh many tons. Even more bizarre is the compressed state of neutron stars, where a single tablespoon of material would weigh an estimated 40 billion tons!

At a certain point of stellar evolution, in response to internal instability, some stars will begin to pulsate in a rhythmic pattern. What is so highly significant about these regular fluctuations is the fact that there is a definite relationship between brightness and period of pulsation. (Thus a Cepheid variable star with a pulsation cycle of 2 days was found to be about 1,000 times as bright as the Sun; while one with a pulsation cycle of little more than a month is some 20,000 times as luminous.) As it is impossible to measure the parallax of a star with any degree of accuracy beyond a distance of a few hundred light-years, and since Cepheid variables are objects of very high luminosity, it will be seen that this particular group of stars has been invaluable in mapping the more distant reaches of the Milky Way, and even to many neighboring exterior systems.

Degenerate Stars

To grasp the sequence of events which link bloated red giants with the crushed state of the white dwarfs, it is necessary to understand what determines the size of a star of given mass. The diameter of a star is actually the outcome of two opposing forces. On the one hand, we have the force of gravitation which tends to draw all matter into a common center of infinite density. In contrast is the explosive or repulsive force of heat – motion which acts to drive the material of a star apart. Condensing out of a hydrogen-rich gas cloud, a newly formed star will cease contracting when sufficient internal heat is generated to support the weight of its overlying layers. At this point it will stop shrinking and become a normal main-sequence star like our Sun, with surface temperature being a reflection of mass – in the sense that large purple-blue stars are the heaviest and hottest and dull red stars are the coolest and smallest. Our Sun is classified as an intermediate yellow star.

Initially, this heat comes from the energy of compression through gravitational contraction. Once the core of a star reaches a critical temperature, nuclear reactions are started which convert hydrogen into helium with the release of a substantial and steady flow of energy. The amount of energy generated in this manner is determined by mass and chemical composition. Of the two, mass plays a truly fundamental role. The reason for this is due to the fact that increased mass leads to higher internal temperatures and to an acceleration in the rate of nuclear reactions which are greatly enhanced by even a modest rise of temperature. This increase in a star's radiation output is out of all proportion to a simple addition of mass. (Thus the giant ultraviolet star Rigel, 30 times heavier than the Sun, shines with a brilliance 40,000 times as great!) As a consequence of this state of affairs the lifetime of a star is very much regulated by the amount of material it contains. However, instead of the more massive stars having more fuel to burn over a longer period of time, it is quite the reverse. Stars of great mass will expend their energy supplies in a comparatively short interval and so meet their ultimate fate in but a fraction of the time needed by the slower burning lightweight stars to exhaust their fuel.

By the middle of the 20th century it became evident that a star will commence expanding as it consumes its supply of hydrogen, very slowly at first and then at an increasing pace as changes in chemical composition and distribution begin to have their effect. As the outer atmosphere of an evolving star expands and becomes cooler and less dense, exactly the opposite conditions prevail deep in the interior. A central core of helium ash is formed and continues to grow in size as more and more hydrogen is burned. Gravitational contraction of the core leads to still higher temperatures and the outer layers of the star are literally blasted further and further apart – causing it to move off the main-sequence en route to the realm of the inflated red giants. Although it has taken our Sun roughly five billion years to evolve to its present state, which is only slightly removed from its original status, one more such interval will lead to profound changes. Red dwarf stars of smaller mass have normal life expectancies measured in tens of billions of years. At the other extreme are heavyweights whose spendthrift nature will cause them to evolve very significantly in less than a hundred million years.

What happens after a critical proportion of a star's hydrogen is consumed? Upon reaching the stage where the generation of nuclear energy is insufficient to support a star's outer regions, expansion is reversed and a degree of gravitational collapse is initiated. As a result of this contraction the star will temporarily become heated and undergo expansion (perhaps ejecting a small portion of its atmosphere) until subsequent cooling permits another contraction. In effect, it may now be transformed into a pulsating star, of which the Cepheids are an example of one particular class or stage of evolution. As such a star continues to shrink, under the influence of gravitation, its dull red surface changes as it passes through successively hotter colors. Contracting far beyond its initial state on the main-sequence it may end up as a small white dwarf star of high density.

Not all stars are privileged to follow this evolutionary loop without interruption, or to end their lives as white dwarfs. Extreme violence lies in store for exceptionally massive stars. Long before a heavy star is able to consume all of its hydrogen, the helium core will reach temperatures conducive to the creation of more complex elements. Helium is converted into such elements as carbon, oxygen, neon, etc. In due course heavier and heavier elements are built up in layers within the star, with the ultimate iron group of heavy metals at the center arising from the burning of the silicon group which had just preceded it. With the build-up of an incredibly dense iron-rich core of atomic nuclei, floating in a virtual sea of electrons, the stage is set for a momentous event. Once this core reaches a critical mass and temperature, electrons are squeezed inside the heavy nuclei where they unite with oppositely charged protons to form neutrons. Occupying less space than that of their combined separate identities, collapse of the core is accelerated – resulting in ever greater numbers of neutrons being formed and an even faster rate of contraction. This eventually leads to a runaway collapse and a catastrophic implosion of the core. The enormous energy suddenly released is sufficient to detonate all of the remaining unburned elements in the rest of the doomed star – producing a monstrous explosion known as a "supernova," in which radiation is released equal to the light of hundreds of millions of suns! Having thus rid itself of much of its mass the stellar remnant is allowed to pursue evolution to its predestined status of a neutron star, or beyond.

As dramatic as a supernova may be, it is not a celestial phenomenon without cosmic purpose. Were it not for these stupendous explosions, which effectively shatter the more massive stars, biological life would have been quite impossible. This is because virtually all of the elements heavier than hydrogen (with the main exception being limited amounts of helium) are produced exclusively in the interiors of stars, especially heavy stars destined as supernova candidates. It is only through explosions of extreme violence that these most vital ingredients manage to become interspersed with clouds of otherwise almost pure hydrogen gas, so that later generations of stars are able to condense out of just the required medium to give birth to planets with the proper chemical composition. Of the matter comprising our very bodies, most of it was once part of a supernova!

The eventual fate of a degenerate star hinges upon the crucial issue of mass, which allows but three choices: white dwarf, neutron star or black hole. Stability is reached, in the case of white dwarf stars, by means of what is termed degenerate electron pressure. In effect, atomic nuclei are able to float contentedly in a sea of tightly packed electrons – from which state it is allowed to very slowly cool over an extended length of time.

The next stage of compression beyond white dwarfdom is that of a neutron star, which is formed when the electron sea is forced inside proton nuclei – producing what is really tantamount to a super atomic particle comprised of neutrons. The mass range of neutron stars is rather limited, with the degenerate neutron pressure stage incapable of restraining gravity much beyond about 2.5 solar masses. Although white dwarf stars are quite plentiful and have been observed for some time, neutron stars are much less common and it was not until 1967 that the first one was finally identified on a photographic plate. Even then, it was as a result of attention being focused upon the origin of strange radio signals from space. A star at the center of the Crab Nebula (an acknowledged supernova remnant that became visible in the year 1054 AD) was found to be flashing on and off some 30 times per second, in perfect synchronization with the radio pulses. Generally referred to as pulsars, such objects are believed to be fast spinning neutron stars embedded in powerful magnetic fields. With pulsation periods commensurate with rotation, they have been described in terms of whirling celestial lighthouse beacons.

The ultimate state of compression is the inevitable fate awaiting degenerate stars possessing masses in excess of what can be supported by this last resort of degenerate neutron pressure. As bizarre as it may seem, the relentless force of gravitation will cause contraction to proceed with ever increasing efficiency – literally squeezing the star out of being! Without a doubt, an implosion of space and time must ensue, producing a situation in which even radiation is unable to escape against the overwhelming force of gravity. A singularity of this order is a *stellar black hole* and may be detected only by its influence on surrounding matter. (The strong X-ray source known as Cygnus X-1, in the immediate vicinity of a star which is cataloged as HDE226868, was the first of a number of stellar black hole candidates to be discovered. Its presence is believed to be revealed through radiation emitted from gases, ejected by its binary companion to form an accretion disk about the black hole and now in the process of being accelerated to enormous velocities as it spirals toward the rapidly whirling infinitesimal core of the singularity.) Essentially, a black hole may be described as an entity that has severed connection with the empirical universe and has entered *an entirely different realm of cosmic existence!*

The Milky Way

Along with many billions of similar companions, our Sun resides in a vast aggregation of stars known as the Milky Way Galaxy. In general shape, our Galaxy bears some resemblance to a fried egg in that its central nucleus is much thicker and more dense than its more outlying regions which tend to trail away at the edges. Through a study of 21-cm radiation, radio telescopes have succeeded in mapping the far reaches of our Milky Way, revealing a basic spiral structure that is so common among the exterior systems. Containing by weight about 1.4×10^{11} (140 billion) solar masses, our Galaxy is probably comprised of something like 400 billion individual stars, extending over a region of the order of 100,000 light-years. With a thickness about one-tenth of its diameter, the lens-shaped main body of our system is enveloped by a spherical halo of more sparsely distributed stars. In every sense, it is truly an enormous concentration of suns.

Like the discovery that our Earth is not at the center of the universe, so astronomers have determined that the Sun is not located within the central bulge of our Galaxy, but is instead positioned between two spiral arms roughly 30,000 light-years from the galactic nucleus. With an orbital motion of close to 150 miles per second, it takes our Sun almost 250 million years to complete one full circuit around the center of our Galaxy. The broad band of encircling light, which may best be observed stretching across the sky on a clear and moonless night, is called the *Milky Way* and owes its appearance to our location near the central plane of our highly flattened system.

In some sections of the Milky Way, the stars are so numerous that their light has blended into what appear to be huge cumulus clouds – clouds which may be resolved into myriad swarms of stars when viewed through a telescope. In addition to these mammoth star fields, our Galaxy is rich in many smaller compact accumulations of stars known as open clusters, of which the Pleides and Hyades clusters are among those in which the brighter members are visible to the naked eye. As a casual glance will show, the Milky Way is not one continuous stream of light but contains conspicuous regions which seem almost devoid of stars. These dark sections are caused by clouds of fine dust lying between us and the otherwise brilliant star fields of the Milky Way. Such clouds serve to scatter and dim the light from the more distant stars, preventing us from viewing what lies beyond. In this manner the central nucleus of our Galaxy is forever hidden to observation by optical instruments.

This enrichment of interstellar gas clouds, through dust from supernovae explosions, has resulted in the existence of two basic types of stellar population: metal-poor and metal-rich stars. The former constitute the oldest stars in our Galaxy (having been born in a primeval environment consisting of almost pure hydrogen with, perhaps, a little helium) and are believed to have ages of at least 15 billion years. The second group of stars are all younger members of our Milky Way and contain a much higher proportion of nature's heavier elements. Situated near the galactic plane, our Sun is a metal-rich star due to its somewhat later birth in a gaseous nebula enriched from earlier generations of supernovae. In contrast, the spherical halo of stars is populated by a preponderance of older metal-poor inhabitants, by reason of a lack of nebulae so essential for the formation of new stars.

Although optical telescopes are unable to penetrate the opaque dust clouds which hide the nucleus of our Galaxy, this is fortunately not the case with regard to radio observation. The picture to emerge is one of considerable interest. An extremely powerful radio source, known as Sagittarius A, is located at the center of the Milky Way. Coming from a region only about 40 light-years across, it is a strong emitter of synchrotron radiation – being produced by the spiraling of high-speed electrons around an intense magnetic field. By far the most logical explanation involves the presence of a *massive rotating black hole* at the center of our Galaxy, possibly of the order of some 5 million solar masses if a recent study of fast-moving gases near the galactic core is any indication. This interpretation is given further support by the discovery of strong gamma ray radiation stemming from this very same region. Uncovered by means of a satellite borne gamma ray detector, this is precisely the generating properties which one might infer of a powerful black hole dynamo.

Forming a concentric halo about the nucleus of our Galaxy, like a swarm of orbiting satellites at distances of some tens of thousands of light-years, are several hundred globular clusters. Typically about a hundred light-years or so in diameter, and containing 100,000 or more suns, they are exceedingly compact aggregations of older, metal-poor, first-generation stars. Featuring many RR Lyrae pulsating variable stars, and almost no uncondensed gas and dust with which to form new stars, their common origin and distance made it easy for astronomers to notice the correlation of intrinsic luminosity with their pulsation period. Also, their distribution about the galactic nucleus served to give man a clear indication of his eccentric location in the Galaxy.

Situated in the Southern Hemisphere, and appearing as two fuzzy patches of light to the naked eye, are twin satellite systems known as the Magellanic Clouds. Classified as small irregular galaxies, both are comprised of many millions of stars and possess a mixed population with regard to age and chemical composition. The Large Magellanic Cloud is about 160,000 light-years distant; while the Small Magellanic Cloud is slightly further away at a distance of 190,000 light-years. Like swarms of smaller globular clusters, they give the impression of being remnants left over from the formation of our Milky Way.

Galaxies of Stars

Galaxies come in a wide assortment of sizes and shapes. Often described in terms of four basic classifications (elliptical, spiral, barred spiral, irregular), further subdivision is capable of giving additional information. Thus a spiral galaxy with tightly wound arms and a rather fat central bulge is called an **Sa** system; one with a moderately large central nucleus and moderately wound arms is defined as an **Sb** galaxy; while loosely wound arms extending from a relatively small nucleus is characteristic of a type **Sc** system. (Our own Milky Way falls into the category of an **Sb** galaxy.) Due chiefly to the fact that there are large numbers of small and dwarf ellipticals, the overall distribution of exterior systems is believed to consist of about 60% elliptical, 20% spiral, 10% barred spiral and 10% irregular galaxies.

In the constellation of Andromeda, at a distance of a little over 2 million light-years (MLY), lies the beautiful spiral galaxy known as M31. This great star system is, in many respects, a twin to our own Galaxy. Similar in both size and shape to the Milky Way, it is encircled by a comparable retinue of globular clusters and possesses a like number of Cepheid variables in its spiral arms. Along with about a dozen and a half smaller systems, mostly of the dwarf elliptical type, these two large galaxies comprise what is called the *Local Group* – a relatively small gravitationally bound cluster some 2.5 MLY or so in diameter.

Moving outward to a distance of almost 7 MLY, we come to another small cluster which is dominated by the spiral galaxy M81. Located in the constellation of Ursa Major, this system could also qualify as a twin to our Galaxy. Other prominent members of this group include NGC 2403, a small spiral system, and the irregular M82 galaxy. Many similar small groupings are to be found as still greater distances are probed. For instance, in the constellation of Canes Venatici, the much photographed "Whirlpool Galaxy" is observed as a major component of yet another weakly-bound gravitational association. Also known as M51, this galaxy is associated with such systems as M63, M101, M94 and NGC 4258, all of which are about 14 MLY distant.

A very noticeable departure takes place, some 50 to 60 million light-years away, in the direction of the constellation of Virgo. Instead of sparse aggregations of galaxies, only a few of which contain really large star systems, an extremely rich cluster is finally encountered. Centered by the huge elliptical (almost spherical) galaxy known as M87, the sprawling Virgo Cloud contains many substantial members and is the closest example of some of the dense swarms which exist in the universe. (Truly a monster among galaxies, M87 is no ordinary system of stars. Not only is it an exceptionally massive galaxy, emitting prodigious amounts of radio waves and X-rays, but its turbulent central core is believed to contain an enormous black hole of the order of some 5 billion solar masses!) This giant Virgo clumping forms the nucleus of a vast assemblage of lesser clusters – including our own Local Group – which is now known as the *Local Supercluster*. Similarly, as the Sun is not centrally located, so our Milky Way is revealed to lie at the outskirts of this overall aggregation.

Typically, a supercluster contains several thousand members and tends to be spread out over a zone of about 150 MLY or more across. With an average separation between supercluster centers focusing upon a figure of roughly half a billion or so light-years, immense voids are seen to be interspersed with such galaxy clumpings. Extending, upon occasion, to distances of at least 300 MLY across, one of these seemingly empty voids could accommodate many thousands of Milky Way galaxies. A subject of growing interest among astronomers, superclusters appear to be distributed in a most remarkably uniform manner when viewed upon the large-scale and very likely constitute the ultimate degree of structuring in the universe.

The Expanding Universe

The principle of the Doppler effect, first expounded by the 19th century physicist Christian Doppler, is crucial to an understanding of what must be considered a truly fundamental feature of the universe. Just as with sound waves, radiation from an approaching source will result in a bunching-up of signals and a shift to shorter wavelengths. Conversely, a receding source will cause radiation to be spread out more thinly and shifted to longer wavelengths. Moreover, for many practical purposes, the degree of this displacement is essentially in proportion to the speed of relative motion – in the sense that the greater the velocity, the greater the wavelength shift. Since the light of a luminous object may be broken down to various reference points, by passing it through the prism of a spectroscopic device, it became feasible to determine speeds of approach or recession simply by measuring the displacement from normal or "rest wavelengths."

The procedure soon became a powerful tool with which to deduce a wide variety of celestial facts, ranging from the motions of stellar binary components to the rotations and movements of entire galaxies of stars. Utilizing this technique, the astronomer Vesto Slipher (of Lowell Observatory) came up with a rather amazing discovery by the year 1925. Inexplicably, some 38 out of the 40 spiral nebulae that he had investigated showed redshifts, which was promptly interpreted as evidence of recession. Furthermore, he had measured speeds as great as 2% of the velocity of light, an incredibly high degree of motion in those days.

The full significance of this peculiarity was not appreciated until 1929, when another astronomer by the name of Edwin Hubble came to supply the missing clue. Having just recently proven that the heretofore mysterious "spiral nebulae" were really distant galaxies of stars, it was noticed that a remarkable relationship existed when the redshifts of galaxies were compared with their distances. Without a doubt, the more remote systems had higher redshifts than the nearby galaxies. In effect, once distances beyond our Local Group were examined, the entire universe appeared to be in a state of expansion! That is to say, except for relatively small gravitationally-bound clusters, the galaxies were all receding from each other – and with speeds evidently in direct proportion to distance. Thus one cluster situated at twice the distance of another was deemed to be rushing away at twice the velocity. Likewise, one at triple the distance was seen to be receding at triple the speed, etc. Known as *Hubble's constant,* this rate of expansion of the universe may be defined as a measure of the time required for a receding system to double its velocity of recession. In terms of philosophical implication, it was to be one of the more important revelations of the century.

Applying this inferred distance/velocity relationship (generally known as the *Hubble law*) to galaxies so remote as to preclude any distance determination by other reliable means, it has been possible to map the brighter cluster members out to distances involving some billions of light-years. Accordingly, a supercluster in Ursa Major, with a redshift of 9,300 miles per second, is receding at about 5% of the velocity of light and is believed to be some 900 MLY away. Another supercluster in Corona Borealis is redshifted to the equivalent of 13,400 miles per second, which is indicative of a recession velocity 7.2% that of light and a distance of roughly 1.3 billion light-years (BLY). In the constellation of Bootes, another supercluster has been found with a redshift implying a recession of 24,000 miles per second, translating to about 13% that of light and a distance of 2.3 BLY. Yet another supercluster, in the constellation of Hydra, is redshifted by 38,000 miles per second, which is equal to a speed 20.4% that of light and a distance of some 3.68 BLY. Upon this basis, the "observable universe" has been acknowledged by many astronomers to have a radius of about 18 BLY, which corresponds to redshift increments of a little more than 10 miles (about 17 km) per second per MLY of separation. At such a distance any velocity of recession is assumed to be equal to the speed of light and, by tradition, would preclude all possibility of viewing what (if anything) lies beyond.

For some decades astrophysicists have been inclined to denote the degree of redshift by the letter "Z." For low velocities, Z is roughly equal to the amount of spectral displacement expressed as a similar fraction of the speed of light. An object redshifted by 5% would be said to possess a redshift of Z = 0.05; while one displaced 10% would be rated as Z = 0.10, etc. This simple, straightforward relationship begins to run into serious problems, however, with contemplation of extreme velocities. In fact, as redshifts of higher and higher values were discovered, it was soon wondered just how long observers could go on breaking previous records. Eventually, spectral shifts were indeed measured which implied velocities of recession well in excess of the speed of light! Realizing the flat impossibility of such a scenario, in terms of traditional physics and established views of radiation propagation, it became necessary for science to introduce so-called "relativistic adjustments" in order to reduce velocities to more plausible levels.

The mathematical form of this adjustment may be stated as follows:

$$1 + Z = \sqrt{\frac{C + V}{C - V}}$$

Where:
 C = velocity of light; and
 V = intrinsic velocity of recession.

Thus a receding source with Z = 0.24 will have its spectral lines displaced by some 24%, Z = 1.00 by 100%, Z = 2.00 by 200%, etc. Upon the basis of formula such examples will be seen to have true velocities of recession of 21.5%, 60.0% and 80.0%, respectively, of the speed of light.

In making this very necessary adjustment of speed and distance, there is now reason to suspect that science has actually managed to adopt a viable formula while approaching the problem from the wrong direction, in spite of using misleading logic in order to bypass the real issue that is involved. Instead of looking to Einstein's theory of relativity for an explanation (and entertaining nebulous ideas of increasing mass and slowing of time with rapid motion), the formula is really no more than a surreptitious reinstatement of the "law of addition of speeds" – a valid law of nature which science supposedly abandoned following an unfortunate misinterpretation of the Michelson-Morley light experiment toward the end of the 19th century. (Of truly immense importance to a proper understanding of numerous cosmological mysteries, this vital issue of light propagation will be discussed at length in *Chapter 3, Cosmological Theories.*)

Quasars

Few discoveries by astronomers have inspired more controversy than detection of the first quasars back in the 1960s. Characterized by a range of unprecedented extreme redshifts, these peculiar starlike objects immediately posed both an enigma and a challenge for man to use them as distance indicators in his quest to map the far reaches of the universe.

The initial mystery hinged upon the realization that quasars are incredibly luminous for their size. If these extreme redshifts were truly indicative of great distances – in the sense that they implied enormous velocities of recession due to expansion of the universe – then their prodigious output of energy seemed incomprehensible in the light of contemporary physics. How could the energy-emitting region of a quasar, perhaps only light-days across, possibly liberate as much energy as a hundred times or more than that of a giant galaxy containing hundreds of billions of stars?

Within a decade or so it was generally agreed that a supermassive black hole must somehow be involved if the inferred energy generation is indeed real. Rotating with extreme rapidity, the dynamo effect of a powerful electric field is combined with the strong gravitational energy of a massive black hole to draw vast swarms of minute matter forms into intimate association. The outcome of such a scenario, according to theory, is a perpetual deluge of newly-created electrons and positrons spewing from the vicinity of a quasar's black hole core. In any event, it is admittedly more than a little ironic that a black hole, from which nothing is expected to escape, should turn out to be responsible for powering the most luminous objects in the universe.

While a huge black hole dynamo might thus solve a pressing energy problem, a few skeptics still doubted whether the major proportion of a quasar's redshift was really due to expansion of the universe. This was to be a matter of controversy for some years to come. An intense gravitational field, it had long been established, could also cause a redshift in spectral lines. The very act of inferring that a massive black hole resides at the center of a quasar was, the skeptics pointed out, an open admission that at least part of their extreme redshifts might be of gravitational origin. (And yet, even this possibility was seen to pose a contradiction since, among other reasons, it would imply an inordinate amount of spectral line broadening that is not in evidence with observation.) To compound the situation, instances were found where some quasars seemed to be associated with galaxies and other quasars of widely differing redshift. About the only decision to emerge, during the early pioneering days of quasar research, was that many of them must surely lie at extreme distances – distances well beyond the confines of our own Local Supercluster.

Over the years evidence steadily mounted in support of what has been termed the cosmological interpretation, which now appears to be quite firmly established in the eyes of most astronomers. Velocity of recession is, first of all, the most logical explanation for such extreme redshifts. Second, the finding of some quasars with spectral line displacements similar to those of certain surrounding galaxies – in apparent compliance with the Hubble law – is strongly suggestive of a common origin and location in space. A third indication arises from the fact that there are no quasars whatever with redshifts indicative of distances of the order of our own supercluster – a clear hint that separations well in excess of several hundred MLY are involved.

On the whole, and conceding instances whereby some quasars do in truth exhibit redshifts requiring a more subtle explanation, it seems highly unlikely that any present confusion must negate all value as potential distance indicators. Typically a hundred times (or more) as luminous as a giant galaxy containing many billions of stars, with some modification of light propagation theory there is now every promise that these exotic objects afford the necessary means by which man may probe to the very edge of the universe.

The Creation Impasse

Traditionally, the most widely accepted explanation for the origin of the universe has been one known as the *Big Bang* model. Inspired by the realization that clusters of exterior systems are all rushing away from each other, it was perhaps inevitable that theory should picture the cosmos as arising from a gigantic explosion. Moving outward from a common center, it was first assumed that all of the elements were formed within the primordial fireball shortly after this momentous event of creation. This hypothesis was soon modified to exclude the formation of atoms heavier than hydrogen and helium, after it was realized that these elements are produced within the interiors of stars. (Also playing a role in this revision was a serious technical difficulty involving the build-up of complex nuclei under prescribed conditions.) According to this concept the impetus of an explosion, which is thought by many to have occurred between 12 to 18 billion years ago, became translated into an expanding universe as clouds of hydrogen atoms condensed locally into individual galaxies and stars.

Certain significant philosophical implications are inherent in this view. First and foremost, it implies that the universe came into being at one specific moment in the past. Hence, at some time in the future, it will have aged to the point where it is no longer able to support biological life. Upon this premise clusters of galaxies will simply continue to recede from each other until, finally, the sky will be void of all exterior systems beyond one's own gravitationally bound cluster. This model is finite in both time and physical constitution. An obvious issue which arises involves the purpose of creation. What motivation lies behind the origin of matter and of the subsequent life forms which it nurtures? On the basis of this view the universe appears highly transitory in essence. Moreover, if God be the sole Creator, where was He before? Did He also come into existence at the same time as the universe? More questions are surely raised than are answered.

From a purely scientific standpoint, this model is beset by serious inconsistencies involving the time factor. In brief, the ages of some celestial objects do not appear compatible with the supposition that the universe was created at one particular instant in the past. While certain portions are unquestionably very old, other regions give every indication of comparatively recent formation. Indeed, even a cursory examination of quasars is sufficient to reveal a contradiction. For instance, since it is generally agreed that the life expectancy of a quasar is much less than the implied age of the universe, one cannot avoid wondering why some nearby quasars are still shining brightly as such if they were all formed about the same time as others whose light would reveal them to be in existence over a dozen billion years ago!

Building upon the explosive origin model, a modified oscillating version was soon proposed and came to be favored by many cosmologists. This concept visualizes the recession of the galaxies as being slowed by gravitation. Upon this premise it is inferred that expansion will ultimately be converted to contraction, so that all matter may be brought together with such force as to produce periodic explosions on a truly cosmic scale. In effect, the universe would be presumed to oscillate between states of expansion and contraction. Although it might appear to eliminate the problem of initial origin, there has been much debate as to whether or not sufficient mass exists with which to produce a closed universe. But a crucial objection to this model is posed by the thought of radiation escaping from the outer extremities of such a universe, for there is simply no way for lost energy to be recovered! To appeal to gravitation is not only useless, but would imply that the weakest field could exert a drastic influence upon the motion of radiation – in itself, a supposition quite opposed to traditional views of radiation propagation. Should just one tiny corpuscle escape, with each cycle of expansion and contraction, in time an oscillating universe would be literally dissipated into space. Accordingly, this concept poses a monstrous contradiction.

In a last desperate attempt to save the Big Bang principle yet another wild modification was proposed: *the Inflationary Big Bang*. After a closer look at the traditional Big Bang model had revealed a host of blatant contradictions with observation, it became very clear that something drastic had to be done. Not only did a remarkably smooth microwave background fail to agree with the large-scale clumping and distribution of galaxies, but the rate of expansion of the universe was insufficient to solve a variety of additional problems which had surfaced over the years. Thus it was theorized that, at the precise instant of the supposed explosion, the initial expansion rate was such that it would actually double in size every 10^{-35} seconds! Following a hundred of these alleged doublings, it was rashly presumed that the universe would, in rather magical fashion, then revert to the rate that is presently observed! Needless to say, this concocted (but widely acclaimed) scenario reminds one of earlier attempts to preserve an incorrect and outworn picture of our Solar System – one in which the Earth was viewed as the central member.

Incidentally, it is to be mentioned that the role that Einstein played in supporting an explosive origin of the universe is most curious. It is, in fact, rather ironic that he chose to retract his "cosmological constant" after the Big Bang model became so popular. (This factor – featuring a "repulsive" side to gravitation – was a later addition to his field equations when it was pointed out that the universe could not remain static, but must either expand or contract if gravitation is propagated indefinitely into space as an attractive force.) In so doing, he probably made the greatest mistake of his career (as we shall see in later chapters). Had he not embraced the Big Bang principle it is doubtful if so many scientists would have been persuaded to follow in his footsteps. In what is surely destined to be recognized as a travesty of science, this blunder by Einstein will be seen to convey a valuable lesson. It is to be hoped that future generations of scientists will soon learn to trust their own faculties of reasoning, instead of being so unduly influenced by the prestige of an esteemed colleague.

An interesting alternative, known as the *Continuous Creation* or *Steady-state* principle, was proposed shortly after the original Big Bang cosmology. This model held that the universe is infinite in time, in the sense that as matter is lost through cosmic expansion and stellar evolution it is replaced by the creation of an equivalence of new matter in the voids between galaxies, so that the overall nature and structure of the universe is preserved. However, the inability to account for the powerful driving force behind cosmic expansion was promptly decried by critics to be an exceedingly weak feature of this model. Discovery of the microwave background (only later to be subjected to valid criticism) was at once declared to be the fading remnant of a fireball announcing the birth of our universe. During the 1960s yet another blow to this model appeared to be delivered by the radio telescope. It was found that the number of fainter radio sources was greater than what might be expected from a simple increase of volume with distance. Since, in looking out into space one is also going back into time, it was inferred that – upon the basis of an explosive origin – man should be able to see the more distant portions of the universe when objects were less scattered through expansion, in apparent agreement with observation.

Now the astonishing conclusion to emerge is that the scientific evidence, as currently interpreted, stands in flat contradiction to all of these cosmological models in some crucial respect. We know that this cannot be so, since one of the two basic creation choices must be essentially correct. (Either the universe always existed or it was created at some point in the past.) Nevertheless, this is exactly the prevailing state of confusion and frustration posed by the study of radiation from distant sources – information enabling man to probe billions of years into the past. The present preference of many astronomers for the Big Bang principle is due not so much to the lack of criticism as it is to seemingly insurmountable objections to the alternate Continuous Creation cosmology.

As a result of this curious impasse, which is reached when one attempts to determine the true cosmological model, it is well to ask whether our dilemma might not arise because of an erroneous assumption with regard to the manner in which radiation is propagated within an expanding universe. Indeed, it is surely more than a mere coincidence that, in almost every instance, the *time* element is somehow directly involved in severe contradiction of the various creation models!

3

Cosmological Theories

Gravitational Attraction – Gravitational Repulsion – Propagation of Radiation – The Construction of Matter – Cosmic Expansion and Energy Conservation

It is true, of course, that science has made considerable progress in uncovering the basic laws whereby gravitation exerts its influence, and also with respect to noting the curious relationship which exists between mass and energy. But in spite of the many experiences which are imparted to us by these forces, it is evident that this knowledge concerns only effects – *the actual root cause or motivation itself remains very much unanswered!* Why, for instance, should any particle of matter have an attraction for another particle? Why does the smallest amount of material contain an astonishing quantity of what might be termed "congealed energy" or "captured motion?" What, in fact, does energy or motion really signify? Proceeding upon the basis of spiritual theory there is reason to suspect that such effects are now capable of a measure of clarification.

Gravitational Attraction

Recalling that the very desire and destiny of all spirit – from the instant of its conception as a lowly "**A**" – is ultimate fusion into the One Undivided Body of God, need we be surprised to observe that a small spark of this inherent desire is manifested by every particle of matter? Essentially, what we choose to call the force of gravitation is none other than an expression of the *desire for advancement* which prevails at the more elementary levels of creation – of the constant striving to overcome an inferior and divided state by means of a literal fusion into fewer entities of higher value. Were we to seek confirmation of the principle of spiritual fusion, from the physical world which surrounds us, *no more graphic illustration could be found than that posed by the phenomenon of gravitation!*

In order to separate unworthy and imperfect levels of spirit, and so prevent an unjustified fusion into One Harmonious Whole, nature has seen fit to introduce the property of distance. Such attributes of the universe as motion and energy are directly related and may, in fact, be looked upon as still another indication of imperfection – or *absence of Total Communion.* It also follows that *rate of communion* is a factor to be considered in conjunction with distance and relative motion. Were we to theorize interaction among components of creation to be instantaneous, for instance, it would be equivalent to the elimination of both distance and time. Speed of communion is thus seen to play a vital role in the overall scheme of cosmic events.

Conceding that gravitational interaction among material particles is actually a manifestation of such spirit's desire for communion (or fusion), the question of propagation becomes of paramount importance. Is this influence restricted to some finite velocity? Or, on the other hand, is it instantaneous in effect, and therefore immune to all events which might otherwise be associated with time? Although quite obviously propagated at an extremely rapid speed, both theoretical and observational evidence appear most implicit in substantiating the view that gravity is not instantaneous. Of all conceivable suggestions as to the velocity of gravitational communion, it would seem that the final solution must center around the *speed of light* – clearly an extremely significant and universal constant of nature.

With this thought in mind, and acknowledging the presence of vast swarms of unseen entities at the lower extremities of the evolutionary pyramid, we may now ask how the influence of gravitation is imposed upon bodies that are both relatively distant and massive. By what specific mechanism are particles of matter able to attract each other? Can we somehow utilize this impressive energy source which is presumed to exist in the form of countless minute quanta? While the basic idea of gravitational quanta – or infinitesimal bundles of energy – being exchanged by material bodies is not altogether new, the mechanics of how this interaction is converted into a means of attraction has hitherto defied explanation.

A vital clue to this ancient riddle very probably resides in what has been termed *Newton's Third Law of Motion:* "To every action there is always an equal and contrary reaction." In short, what is implied in the case of gravitational attraction is that, whatever communion a body may receive from a source outside of itself, so it must promptly emit or relinquish an equivalence of such energy. (Were a material particle able to retain every quantum it encounters it would tend to grow infinitely greater.)

As a prelude to an attempt to resolve this problem, and giving due recognition to the principle of a constant exchange of energy among material structures, certain postulates may be stated as follows:

1) Gravitational quanta differ fundamentally from other material particles only in size or status and may be described as elementary forms of matter which have yet to evolve – through the process of fusion – into more advanced structures. Consequently, in compliance with the premise that relative abundance is a factor of intrinsic worth, a wide variation in the status or mass of such quanta is bound to exist. That is to say, the lower the status the more abundant the structure.

2) Far from being instantaneous in effect, the propagation of such quanta is restricted to some finite velocity – essentially a speed strongly focusing upon that of light itself. It may be presumed, in fact, that light speed represents *"escape velocity"* of infinitesimal quanta from association with some larger structure.

3) In general, when a typical material particle receives a quantum of gravitational radiation it will be unable to retain this additional energy and must quickly emit any surplus. Moreover, unless prevented by some infrequent occurrence, there will exist a pronounced tendency to eject this surplus in the *opposite* direction from whence it was received. In this event, the inferred direction of emission is likely facilitated by the rapid rotation of the attracting body. Unable to penetrate the dense concentration of energy which constitutes a major atomic particle, a tiny gravitational projectile will be momentarily carried in the flow of this overall momentum. Finally, after rotation of the principal body has proceeded to the point where it no longer poses an obstruction, the newly acquired quantum is permitted to sever its association and resume its previous course with a final escape velocity equal to that originally possessed: the speed of light.

4) As infinitesimal quanta approach within a critical distance of a material particle, they will tend to come under the influence of an

intangible force which might well be defined as anti-chance. The end result is a greater influx of energy in the immediate vicinity of the material body, so that more quanta are received in unit time than could be explained on grounds of chance encounter alone. Such dynamic intervention may be ascribed, in principle, to the inherent desire for fusion which pervades all creation. As a consequence of this inborn desire, every particle of matter may be visualized as possessing a "halo" in the form of an encircling cloud of energy – an atmosphere of unassimilated and relatively minute quanta which is prevented from obtaining an unconditional escape due to the constant striving of spirit to secure fusion. Upon this basis, it will follow that all halo characteristics must be determined somewhat by the mass of the central body, in the sense that a more massive nucleus will exhibit a more extensive halo and therefore be capable of retaining larger quanta.

Armed with the assumption that infinitesimal quanta are forever being absorbed and emitted by all material bodies, the task remains to relate gravitation with this exchange of energy. How is it possible to achieve attraction through innumerable encounters with these tiny quanta, acting as they must like a swarm of miniature projectiles?

Predictably, initial reaction to a larger-than-average flow of energy from one particular location in space – the direction of the most dominant mass – will be a slight recoil away from the point of impact, as high speed collisions serve to drive the body backward. Within the smallest fraction of a second, however, this motion of recession is terminated by the expulsion – in the *opposite* direction from whence it was received – of the excess energy just obtained. Acting upon the principle of jet or rocket propulsion, this ejection will quickly return the particle to its original position. At this point, the attracted mass will be made to oscillate rapidly with each cycle of absorption and emission, with the most conspicuous motion occurring along the plane of greatest interaction. The degree of this wavelike motion would be largely determined by the mass of the particle so involved, in response to the logical premise that the more massive the body the less inclined it will be toward oscillation.

The key to the actual process, whereby small-scale oscillation is transformed into large-scale motion of attraction, would appear to have a solution in the production of what might well be described as *"neutral quanta."* Just as swarms of infinitesimal projectiles may collide with bodies of much higher mass, so in regions of pronounced interaction numerous collisions must also occur among the tiny quanta themselves. Following a collision of two such quanta, there is bound to prevail a brief instant when their respective motions through space are retarded or neutralized to a significant extent, or even canceled altogether as in the case of head-on collisions among those quanta of similar size. Considering that impacts of this type will be far more frequent in regions of concentrated mass, and especially along the plane of most intense interaction or exchange of energy, it will be inferred that therein exists a substantial supply of "neutral quanta" which an attracted body may temporarily absorb – without penalty of incurring a backward thrust in proportion to the energy received – and thence to utilize it as "fuel" by reason of its subsequent ejection at the velocity of light. Thus the mechanism of gravitational attraction could well be explained through the acquisition of "neutral quanta" – *energy which may be absorbed into a system with reduced or negligible impact, but which is later ejected in the opposite direction at high velocity and with a recoil action sufficient to propel the particle in the direction of this source of energy!*

One important aspect of this proposal demands clarification. As an attracted particle is made to oscillate, through absorption and emission of energy, it will follow that the strongest concentration of potential "fuel" must tend to lie along a narrow plane extending from the attracting mass straight through the center of the attracted body. But since the combined flow of outgoing and incoming energy will be the same on opposite sides of an attracted particle, it may be deduced that the frequency of collisions will be such as to produce as much "neutral quanta" on the side opposite the attracting mass as on the reverse side. Obviously, at this point, a successful interpretation of acceleration must hinge upon the acquisition of a larger proportion of this "fuel" from the direction facing the attracting mass. Unless this can be achieved, the theory will collapse of its own accord.

Invariably, this preferred assimilation is facilitated by the general flow of energy itself – a stream which flows most intensely in a direction from the attracting body toward the attracted particle. For one effect of this predominant stream must be to create a certain proportion of *secondary* collisions, as "neutral quanta" produced by earlier impacts are in turn struck by other members of this stream before they have had time to fully recover their momentum. Of truly fundamental importance, however, is the realization that when a secondary collision occurs between two interacting quanta the tendency will be to sweep this energy *toward* the attracted body. On the other hand, when such collisions are produced on the opposite side of the attracted particle, this potential source of "fuel" will tend to be swept away by the force of the prevailing stream, since the direction of flow is away from the interacting members. As a result, a high percentage of "neutral quanta" will be absorbed from the direction immediately facing the attracting mass. With the subsequent ejection of this "fuel" – in a direction *opposite* to that received – attracted bodies are thus enabled to propel themselves toward any influencing concentration of mass.

It will now be seen why an adjustment to Newtonian mechanics is required in order to express the factor of motion – simply because any external force must reflect the inability of gravitation to propagate at full efficiency when communion is limited to the velocity of light. However, rather than be misled by traditional relativistic notions of increasing mass with extreme motion, it will be acknowledged that it is really a case of *declining gravitational efficiency with increased relative motion,* as such movement makes it harder for quanta to catch up to rapidly receding bodies! Accordingly, at the speed of light itself the influence of gravitation must be virtually equal to zero. The popular but highly nebulous idea of mass becoming infinite at the speed of light is both illogical and contradictory. Indeed, radiation is propagated at just this speed and nobody has ever claimed infinite mass! A modified relativistic formula, depicting a reduction in the efficiency of gravitation with increased relative motion, is given in Figure 2.

Gravitational Repulsion

In actuality, *not all infinitesimal quanta may be exchanged in such a manner as to induce attraction!* As opposed to traditional belief, gravitation can no longer be viewed as an attractive force that propagates indefinitely into space – one which merely "varies directly as the product of mass and inversely as the square of the distance." For while this classical law is essentially valid within our Solar System, and may be applied with some measure of confidence to galactic phenomena in general, there is every reason to suspect that this is not the case when extremely large distances (such as those which exist between clusters of exterior systems) are considered.

According to theory, the susceptibility of infinitesimal quanta to the influence of material particles hinges strongly upon the issue of size. In other words, smaller quanta will pose less of a challenge to a spiritual body attempting to exert its will – or desire for fusion – over other members of creation. (It is for this reason that the halos enveloping "maximum" particles, of the order of electrons and protons, are believed to be comprised of the more elementary levels of quanta.) Larger corpuscles must increase the task of securing association and will, therefore, be more likely to incur

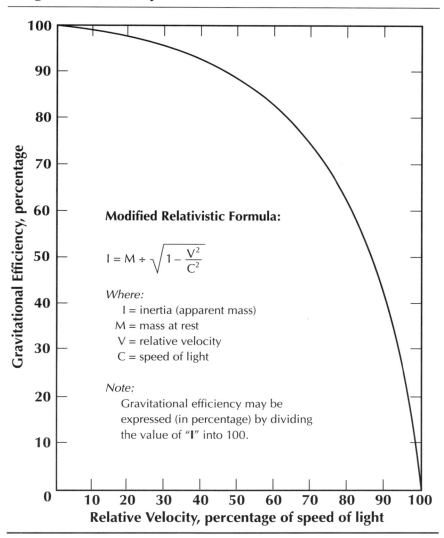

Figure 2. **Efficiency of Gravitation with Relative Motion**

Modified Relativistic Formula:

$$I = M \div \sqrt{1 - \frac{V^2}{C^2}}$$

Where:
 I = inertia (apparent mass)
 M = mass at rest
 V = relative velocity
 C = speed of light

Note:
 Gravitational efficiency may be expressed (in percentage) by dividing the value of "I" into 100.

rejection. Nevertheless, quanta too large to permit halo membership may be induced to follow trajectories of some curvature, in the sense that the more substantial the quanta the less affected will be their paths. Applying this line of reasoning to the structure of an entire galaxy, it means that the vast bulk of exchanged energy is caused to remain in circulation within the confines of that system.

But while the majority of ejected energy is unable to escape into the depths of outer space, it is also clear that this restriction does not apply to the more massive quanta. In fact, it is most logical to assume that a significant proportion of quanta – residing below the spectrum of light and radio emission – is indeed able to pass freely into space. Very conceivably, it could be the manner in which these larger quanta – escaping from clusters of galaxies – interact with the material particles of other systems *which holds a vital key to the observed expansion of the universe!*

Inasmuch as quanta of greater mass will, through sheer force, be capable of penetrating deeper into the halos of "maximum" particles – and so encounter regions of higher density or energy content – the end result is very likely to be a prompt rejection and scattering of this surplus, particularly toward the direction from whence it came. In effect, there will arise a situation whereby spiritual bodies are suddenly called upon to "process" more energy than they are capable of handling. As an injustice of this nature becomes more extreme, so it could demand a more immediate expulsion of such foreign quanta – quite regardless of whether they have had sufficient time to be whirled around to the opposite side of the principal body. Hence, in response to an injustice perpetrated against both host and invading quantum alike, an incentive will exist for the foreign body to quickly sever all bonds and rebound much as a ball will bounce off a hard surface.

We now see how an exchange of energy, among the various clusters of galaxies, could lead to mutual repulsion and an expanding universe. For in the process of exchanging quanta, of a certain mass range, two distinct instances of repulsion will be inferred as this impacting energy interacts with the major particles of such systems. Beyond the initial impact, which is a force directed away from the source, there will be a secondary repulsive thrust as this undeserved quanta is swiftly rejected and expelled in the general direction from whence it arrived.

Since only a relatively small proportion of all exchanged quanta will so interact as to induce repulsion, it follows that any inherent repulsive characteristics will be entirely masked by the predominant tendency to induce a state of "localized" attraction. On the whole, this universal exchange of energy may be viewed as the simultaneous existence of two opposing forces – with the force of attraction simply being the stronger at more moderate distances. With contemplation of the separations between clusters of galaxies, the hitherto dormant factor of repulsion is finally permitted to gain the upper hand, as the only quanta capable of being exchanged over such distances are the larger corpuscles which must act to induce repulsion.

This "dual" concept of gravitation, incidentally, seems to be just what is needed to resolve a curious mass discrepancy involving clusters of galaxies. Astronomers of the 20th century were puzzled to find large inconsistencies in calculating the total masses of such associations by two proven methods. One procedure, known as the "virial theorem" technique, depends on measuring the motions of member galaxies and relating this information with the principle that speed – within a gravitationally bound system – is a factor of mass. Alternatively, by computing the individual masses of the star systems comprising a cluster, simple addition will yield a value. Invariably, a highly disconcerting discrepancy arises upon a comparison of the two methods. It was found that the kinetic energy of the galaxies implied masses far beyond that revealed by a more detailed determination of separate masses. Moreover, it was discovered that the larger the cluster the greater this "surplus" motion which seemed to demand interpretation as excessive overall mass. This strange paradox has persisted in spite of careful investigation – presumably, serving notice that something may be going seriously awry when traditional views of gravitation are extrapolated to really large distances!

Recognition of a long-range repulsive factor, arising as a result of a universal exchange of minute quanta, affords a most plausible explanation of this dilemma. At distances which separate cluster members, the usual overwhelming aspect of gravitational attraction is greatly weakened. Accordingly, while still overshadowed, the inferred repulsive side of this energy exchange is permitted clear expression. There will exist, in fact, a pronounced tendency for the constituents of a cluster to push each other apart. Prevented from doing so by a superior binding force, this energy must reveal itself in terms of enhanced galactic motions. In short, it is simply translated into the "surplus" kinetic energy which appears to undermine any overall mass determination. It is surely no accident that the magnitude of this discrepancy, which is generally a factor of 10 to 20, reflects the size differential between the diameters of the galaxies and the actual distances separating cluster members!

Upon the basis of a finite physical universe – and only a universe of finite dimensions – this underlying repulsive influence will be permitted an opportunity to express itself in rather dramatic fashion. Since nothing will be held to exist beyond the outer boundary of the universe which would exert an inward pressure upon nearby clusters of galaxies, it may be deduced that the outermost clusters will be free to utilize this repulsive force to the fullest. Moreover, once started, any outward movement must be cumulative in effect, in the sense that a recession of the outlying systems will reduce any inward pressure upon the more central members – thereby allowing them to share this large-scale motion without any loss of efficiency. Thus the universe will be caused to expand at a constant rate, in full agreement with the Hubble distance/velocity relationship which holds that the speed of recession is a direct factor of distance; that is, a twofold increase in separation will be reflected in terms of a twofold acceleration in velocity of recession.

Yet another indication of this "dual" nature of gravitation may well be found much closer to home – in fact, within the confines of our own Milky Way galaxy. In conformity to Newtonian laws of gravitation, the outer portions of a galaxy should rotate much slower than the innermost regions with their high concentrations of mass. To the great surprise of astronomers, it was discovered that the outlying sections of all spiral and elliptical galaxies rotate considerably faster than expected. Why? Unwilling to acknowledge a repulsive side to gravitation, they are compelled to postulate large quantities of unseen mass in the extremities of these galaxies. What form this "missing mass" might take has continued to remain a big mystery, in spite of an abundance of wild speculation. Hopefully, it will not be long before it is finally realized that the issue is really one of "missing physics," in which the repulsive side of gravitation is conceded to supply the missing rotational impetus.

Still closer to home we have such anomalies as perihelion (point of closest approach) advance of the planet Mercury, along with orbital peculiarities involving Uranus and Neptune. While most scientists are inclined to believe that Mercury has a solution in Einstein's famous relativity theory, it is to be noted that a viable alternative now exists. Paradoxically, gravitational repulsion can serve to *both advance and retard perihelion!* In the process of exerting a "push" away from an interacting body, this proposed force will reduce the strength of attraction – imposing perihelion advance. On the other hand, the very action of imparting thrust must also increase orbital distance and extend transit time – thereby retarding any advance of perihelion. The problem is therefore seen to be one of assessing the relative value of the two opposing forces. (The mathematics of this relationship is given by the author in his book *The Eternal Universe,* where both Mercury and a binary star system known as DI Herculis are revealed to share a similar distribution of opposing energy flows. Likewise, perturbations in the orbits of the outer members of our Solar System are shown to have an explanation in the principle of gravitational repulsion.)

Yet another instance of orbital anomalies may well involve a large-scale effect associated with the formation (and preservation) of spiral structure in galaxies. Considering that the extremities of all elliptical systems take much longer to complete a revolution about a galactic center than do the innermost regions, it has been wondered how spiral arms could persist after a few orbits. The very fact that such structure is so common is a clear indication that some subtle force is at work upon a long-term basis. By far the most promising theory to be advanced is a density-wave hypothesis, in which rotating shock waves play a major role in compressing gas and dust into regions of new star formation – producing luminous spiral arms. An essential requirement of this scenario is a pattern whereby matter is assumed to describe elliptical orbits about a galactic nucleus, with successively smaller orbits offset, relative to outer orbits, in the same direction as the overall rotation of the system. A lingering problem, however, has been one of defining the actual mechanism responsible for this prescribed systematic alignment of orbits. In essence, we are likely dealing with exactly the same set of circumstances responsible for the perihelion advance of Mercury. Although upon a vastly larger scale, there is every reason to suspect that the existence of galactic spiral structure is an indirect manifestation of gravitational repulsion.

Upon due consideration of a possible formula with which to define this repulsive aspect of gravitation, it seems highly probable that the solution lies in ascribing a *twofold increase of repulsion for every fourfold increase of distance!* Radiating from the immediate vicinity of elementary particle nuclei, as a negative force some 10^{20} times weaker than the positive influence of gravitation, it may be presumed to increase with separation as stated – becoming equal to that of attraction at a distance of roughly 17 million light-years (MLY). What is so very remarkable about this distance/repulsion (D/R) factor (see Figure 3) is the excellent agreement of calculations with observation at such test distances as those which charac-

Figure 3. Distance/Repulsion Relationship of Gravitation

Distance	Repulsion	Distance	Repulsion
Fourfold increase (from elementary particle nucleus)	**Twofold increase** (in terms of gravitational attraction)	**Fourfold increase** (from elementary particle nucleus)	**Twofold increase** (in terms of gravitational attraction)
1.5×10^{-13} cm	10^{-20}	110.680 km	8.590×10^{-10}
6.0×10^{-13} cm	2.0×10^{-19}	442.722 km	1.718×10^{-9}
2.4×10^{-12} cm	4.0×10^{-19}	1,770.89 km	3.436×10^{-9}
9.6×10^{-12} cm	8.0×10^{-19}	7,083.55 km	6.872×10^{-9}
3.84×10^{-11} cm	1.6×10^{-18}	28,334.2 km	1.374×10^{-8}
1.536×10^{-10} cm	3.2×10^{-18}	113,337.0 km	2.749×10^{-8}
6.144×10^{-10} cm	6.4×10^{-18}	453,347.0 km	5.498×10^{-8}
2.458×10^{-9} cm	1.28×10^{-17}	1.813×10^{6} km	1.100×10^{-7}
9.830×10^{-9} cm	2.56×10^{-17}	7.254×10^{6} km	2.200×10^{-7}
3.932×10^{-8} cm	5.12×10^{-17}	2.901×10^{7} km	4.398×10^{-7}
1.573×10^{-7} cm	1.024×10^{-16}	1.161×10^{8} km	8.796×10^{-7}
6.291×10^{-7} cm	2.084×10^{-16}	4.642×10^{8} km	1.759×10^{-6}
2.517×10^{-6} cm	4.096×10^{-16}	1.857×10^{9} km	3.518×10^{-6}
1.007×10^{-5} cm	8.192×10^{-16}	7.428×10^{9} km	7.037×10^{-6}
4.027×10^{-5} cm	1.638×10^{-15}	2.971×10^{10} km	1.407×10^{-5}
1.611×10^{-4} cm	3.277×10^{-15}	1.188×10^{11} km	2.815×10^{-5}
6.442×10^{-4} cm	6.554×10^{-15}	4.754×10^{11} km	5.629×10^{-5}
2.577×10^{-4} cm	1.311×10^{-14}	1.901×10^{12} km	1.126×10^{-4}
0.010308 cm	2.621×10^{-14}	7.606×10^{12} km	2.252×10^{-4}
0.41232 cm	5.243×10^{-14}	3.215847 LY	4.504×10^{-4}
0.164927 cm	1.049×10^{-13}	12.86339 LY	9.007×10^{-4}
0.659707 cm	2.097×10^{-13}	51.45355 LY	1.801×10^{-3}
2.63883 cm	4.194×10^{-13}	205.8142 LY	3.603×10^{-3}
10.5553 cm	8.389×10^{-13}	823.2568 LY	7.206×10^{-3}
42.2212 cm	1.678×10^{-12}	3,293.03 LY	0.01441
168.885 cm	3.355×10^{-12}	13,172.1 LY	0.02882
675.540 cm	6.711×10^{-12}	52,688.4 LY	0.05765
2,702.2 cm	1.342×10^{-11}	210,754.0 LY	0.11529
10,809 cm	2.684×10^{-11}	843,015.0 LY	0.23058
43,235 cm	5.369×10^{-11}	3,372,060.0 LY	0.46117
1.72938 km	1.074×10^{-10}	13,488,239.0 LY	0.92234
6.91753 km	2.147×10^{-10}	(16,900,000.0 LY)	(1.0000)
27.6701 km	4.295×10^{-10}	–	–

terize galactic cluster members, galactic extremities and halos, and the outer constituents of our Solar System. (At the distance of the planet Neptune from the Sun the inferred repulsion is about 193,000 times weaker than the attraction.) In every instance there would appear to be strong confirmation of the implied degree of repulsion.

Additional evidence in support of this D/R aspect of gravitation may be found in the "bubble-like" arrangement of galaxies that has now become apparent.* Contrary to widespread belief, which would hold gravitation to be solely an attractive force, most galaxies are so distributed as to suggest an opposing influence; for they tend to reside toward the edges of "shells" or "bubbles" – with near voids at their centers. *This is precisely the pattern to be expected should gravitation manifest a "dual" nature!* Indeed, this observation is quite incomprehensible in terms of traditional cluster dynamics, which should feature central condensations of mass – not the reverse state of affairs with such regions almost devoid of galaxies!

Propagation of Radiation

A number of revolutionary conclusions must arise as a result of our new interpretation of gravitation. The mysterious "wave" aspect of radiation, for instance, promises to make some sense. We can readily understand why material particles exhibit wavelike characteristics, as they are caused to oscillate through periodic absorption and emission of minute quanta. But can we adopt this same principle to account for the rhythmic wave movement as it streaks through space? Considering such a phenomenon as the polarization of light, it soon becomes quite evident that a more subtle explanation is in order.

By far the most logical solution would seem to involve the view that radiation does not normally consist of solitary particles. Just as many stars are members of double systems, so it may be expedient to visualize radiation as miniature *binary systems* – essentially, two entities revolving about a common center of interaction while moving through space as a unit. (Polarized light may be pictured as swarms of minute binary systems possessing a specific and common orientation in space.) Due to this rotary motion, radiation will be observed to display a spiral or "wave" trajectory,

---

* "A Slice of the Universe," *The Astrophysical Journal Letters,* March 1, 1986.

with the particular wavelength itself simply requiring interpretation as an expression of the degree of separation of the binary components.

It may well be wondered why science ever hesitated to adopt this simplistic wave-particle description of radiation. The most obvious objection is one which threatens to ridicule the very suggestion when we contemplate radiation of longer wavelengths, since it involves the lack of an attractive force sufficient to retain the structure of such weak systems. Inasmuch as most binary components would have to be assigned separations exceedingly vast by nuclear standards, it may be asked why these associations are not immediately disrupted by much closer encounters with far more massive and influential particles.

Upon the basis of a limiting velocity on behalf of gravitational interaction (e.g., the speed of light) and with insight into how acceleration is accomplished, it will be seen that by the time full communion could be established with any nearby material particle a tiny bundle of radiation will have left that particular region of space and will, therefore, be largely unaffected. (Invariably, an exceptionally close encounter or collision would be another matter.) Thus, by reason of a common motion through space which would permit a state of mutual communion to exist, a binary system of quite minimal gravitational attraction may be constructed – *as long as this system possesses the velocity of light relative to any body which might tend to disturb it!*

At this point it is necessary to ponder certain implications of the term "velocity." A fundamental scientific premise is that no material body or corpuscle of energy can exceed the speed of light. But it is at once evident that a body possessing this "maximum" or limiting velocity relative to one system will not possess this same speed with respect to all other moving systems. Two light photos traveling in opposite directions will obviously exceed the speed of light relative to each other. Likewise, clusters of galaxies clearly exist which display combined velocities of recession in excess of light speed. Hence, in a sense, the velocity of light can be surpassed and any inferred restriction very much remains to be clarified.

Essentially, the speed of a body through the medium of space is regulated by its acquisition of "neutral quanta" – with acceleration proceeding just as long as it is possible to obtain and utilize such "fuel." Should we theorize motion in excess of the speed of light, energy encountered in the direction of flight will strike with greater force than the recoil action with which it is later ejected – thereby constituting a braking action. On the other hand, if a body is moving close to light speed, relative to the prevailing gravitational field, it is evident that only a minimal amount of quanta could ever manage to catch up to it from behind in order to produce "neutral quanta." It is for such reasons that velocity becomes so critical at extreme speeds, and which accounts for the self-terminating aspect of motion once the velocity of light is attained relative to the prevailing gravitational influence.

In assessing all aspects of acceleration, it will be noted that any adjustment of speed is the product of both velocity and density of the medium through which passage is attempted. In a sense, extended motion through a tenuous gravitational field is really equivalent to a shorter encounter with an intense field. Nevertheless, it is most essential to understand that it is the actual velocity of a body which becomes of overwhelming importance at extreme speeds. For as light speed is approached, the chief effect of a stronger gravitational field will be to merely expedite any required adjustment of velocity – *an adjustment which the weakest field will surely facilitate, even if it should take a slightly longer period of time!*

Certain consequences of extreme motion warrant discussion. It may be inferred, first of all, that a body moving through a gravitational medium at close to the velocity of light will exhibit an increase of mass due to the greater volume of quanta swept up and momentarily detained, where it must add to the strength of the halo. But rather than be misled by a popular misconception, which would visualize mass as being increased by an indefinite amount once the velocity of light is attained by a material body – an obvious impossibility – it will simply be conceded that no further acceleration is possible due to the inability to secure propulsive "fuel."

Seemingly, a notable feature of rapid motion is the curious effect of time retardation – a phenomenon which is believed to have received a measure of verification through laboratory experiment, even if the mechanics involved have been far from understood. It has been found, for example, that the time required for certain unstable atomic nuclei to undergo spontaneous disintegration is a factor of velocity. Having little influence at low speeds, as the velocity of light is approached this retardation effect becomes quite pronounced. It may be wondered how an internal instability problem could be affected by such a factor as external motion. What happens, of course, is that increased speed leads to the acquisition of a more substantial halo – an encircling swarm of minute quanta which serves to insulate the nuclear body from the outside world. But since insulation works in both directions, it follows that there will exist a tendency to preserve whatever resides within the halo. In order for disrupted components to escape, they must first break through the enveloping atmosphere of the halo, a feat which will become somewhat more difficult should the strength of the halo be enhanced due to increased motion within a gravitational field.

From a philosophical basis, it may be concluded that the velocity of light represents a typical expulsion from an undeserved level of fusion, or spiritual association – *motion which is relative to the entity it has just left!* In essence, the "escape velocity" imparted to expelled quanta is an expression of *unworthiness,* of complete and total separation with respect to the system with which it was so unjustly and momentarily associated. Applied to the propagation of energy through space this means that *whatever field of influence a tiny bundle of radiation may enter, so it will be caused to travel with the velocity of light relative to that gravitational system!*

It becomes possible, at this point, to gain much needed insight into a mystery which was raised by the classical experiment pioneered by Michelson and Morley toward the end of the 19th century. Up until this time it was firmly believed that the propagation of light would obey the same fundamental laws of motion as does the material bodies with which we are so familiar. For instance, if we expel a projectile from a moving object – in the direction of its motion – it will be seen to inherit this initial velocity in addition to any later thrust. This phenomenon is known as the "law of addition of speeds." It was therefore expected that light quanta, emitted from moving sources and striking other moving objects, would display variations in velocity commensurate with such movement.

The results of experiment, however, were most conclusive and left no doubt as to the fact that something very strange must transpire with respect to tiny photons moving with extreme velocity. It was found that radiation is propagated at what appears to be a "constant" velocity; for regardless of the motions of either source or receiver the speed of impact was the same. In effect, it was as though some intangible force had intervened and accelerated any photons which approached at less than a certain speed and resisted or slowed the motion of those approaching in excess of this same limiting velocity. The nature of this mysterious intervention may now be defined, of course, in terms of absorption and emission of energy by radiation passing through the swarms of infinitesimal quanta so permeating space and constituting a gravitational field – essentially, a process which assures propagation at the speed of light relative to the prevailing influence.

COSMOLOGICAL THEORIES

Unfortunately, the Michelson-Morley experiment received a false and rather curious interpretation which is destined to go down in history as one of the greatest blunders of modern science. It was actually assumed that when a light photon is emitted in our direction, from a receding celestial object, it will at once begin to approach our system with an effective speed of some 186,000 miles per second – while yet exhibiting an identical velocity of recession with regard to the object which it has just left. Initial relative motion – between source and receiver – is totally ignored! Accordingly, the distance of a receding source (in light-years) is also thought to be a measure of the time (in years) that has elapsed since this light first began its journey. But this view involves unwarranted and contradictory postulates. Not only does it necessitate an unexplained break with the "law of addition of speeds," but it relies upon sheer magic to be able to account for its miraculous ability to approach all bodies situated in its line of flight with a speed of 186,000 miles per second – irrespective of whether such bodies are in fact approaching or receding!

Curiously, without realizing the full implications of traditional redshift adjustments, astronomers have been using a calculation which actually takes into consideration the perfectly valid "law of addition of speeds" – a law that was ostensibly negated upon acceptance of the theory of relativity! Ironically, in the process of translating all extreme redshifts to more reasonable values, they have inadvertently stumbled upon the very formula which is capable of describing the time of light propagation from distant receding sources. (This so-called relativistic adjustment of redshifts is discussed in *Chapter 2, Our Physical Universe,* in the section entitled "The Expanding Universe.") Instead of merely a means of defining a degree of spectral displacement, the adopted Z term is really a measure of the *time required for radiation to travel from source to detection within the confines of an expanding universe, and in full compliance with the "law of addition of speeds."* (See the curve in Figure 4.)

All that is necessary to transform a Z redshift into the true time of light propagation is to multiply it by the radius of the universe (the point where recession equals the speed of light) as expressed in light-years – a figure which will always be some 1-1/2 times that of Hubble's constant. The reason for this, of course, is due to the fact that velocity (relative to the center of the universe) is accelerating steadily with distance. For example, at a point equidistant from the center and edge of a universe with an 18 billion light-year (BLY) radius, a source would be receding at 50% of the speed of light. In order to double both velocity and distance, in one cycle of Hubble's constant, it is necessary to travel a further 9 BLY and arrive at the edge of the universe with a speed equal to that of light. Since the average motion of such recession will be 75% of the velocity of light, it may be calculated that the time required will be of the order of 12 billion years. Hence, Hubble's constant will be 12 billion years in a universe having a radius of 18 BLY. (See the diagram in Figure 5.)

Figure 4. Redshift/Time/Velocity/Distance Scale *(Concentric View)*

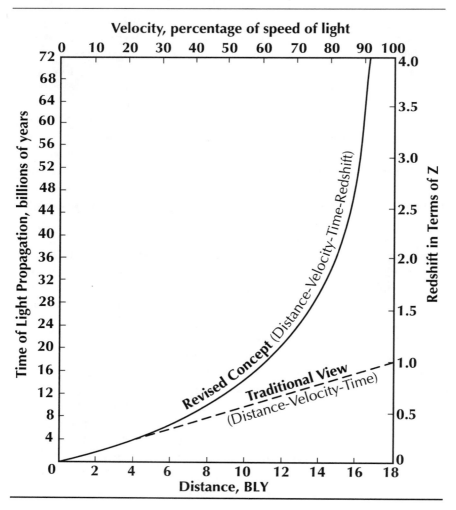

Figure 5. **Expansion of Universe and Propagation of Light**

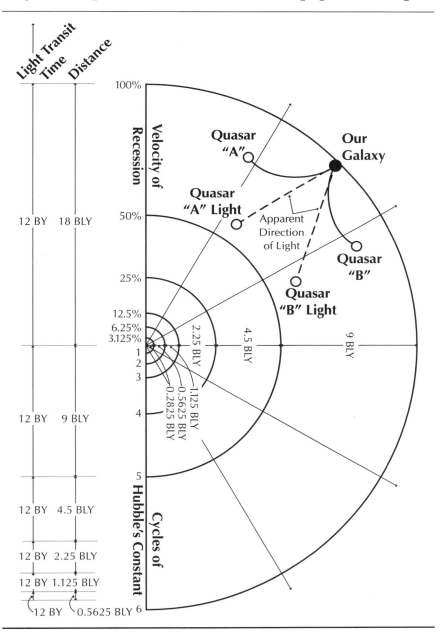

In actuality, the time of light transit from many distant receding sources will be enhanced somewhat by a rather unique and hitherto unsuspected curvature of space. Due to expansion of the universe and the steady outflowing of all forms of mass/energy, there must prevail a perpetual influence tending to push radiation toward the nearest edge of the universe. The end result of this outward streaming of space, will be the bending of radiation received from all directions not in harmony with this overall flow. In effect, *there will appear to be a general deflection of images toward the center of the universe* – a displacement which must include all extragalactic sources not in our line of sight with either the central core or the closest boundary of the universe!

The degree of this "arcing" (or bending of the path of radiation), from remote sources partaking in the expansion of the universe, will hinge upon three factors. First, it is an effect *which is enhanced by increased distance.* Second, while incapable of imposing "arcing" upon radiation pursuing trajectories in harmony with the mass outward flowing of space, *it will be most pronounced at right angles to this overall movement.* (See Figure 5.) Finally, such bending of radiation is magnified by *proximity to the edge of the universe* – in the sense that extreme curvature must arise and become apparent to an observer close to the cosmic edge, where expansion velocity approaches the speed of light itself.

What is so promising about this new view of radiation propagation is its ability to resolve the creation impasse. It will now be seen that redshift measurements of distant celestial phenomena must reflect both actual distance – at time of emission – and the distance by which the universe has expanded during the interval that has elapsed while in transit. Essentially, *such redshifts are a measure of the time that has elapsed since radiation emission!* Thus there will exist a discrepancy in any time/distance determination between the traditional viewpoint and our revised interpretation – a difference that is almost negligible for neighboring objects, but which becomes of overwhelming importance when more remote sources are considered. (See Figure 4.) In short, due to reinstatement of the "law of addition of speeds," it follows that radiation – from a distant receding source – will spend the vast proportion of its journey toward us with an approach velocity that is somewhat less than the speed of light!

By virtue of this revised concept, it finally becomes possible to refute evidence which many astronomers had accepted as conclusive in eliminating the principle of Steady-state cosmology. Hitherto, the discovery of an excessive number of faint extragalactic radio sources (beyond what might be expected from a simple increase of space/volume with distance) seemed to support the Big Bang model. However, we now see that – at high redshifts – we are actually overcoming *more time than distance!* Rather than detecting larger numbers of remote sources in a smaller volume of space, it is seen to be a case of observing more objects of a significant age group. In effect, *we are able to view substantially more than one generation of creation!*

Should we accept this proposed revision of radiation propagation, which implies that we have already detected objects with light transit times some tens of billions of years greater than what is permissible upon the basis of any explosive origin of the universe, *then there is no alternative but to conclude that the highly dynamic principle of Continuous Creation emerges in sole agreement with the evidence!*

What were once considered impressive objections to Steady-state cosmology are thus revealed to be without foundation. But the very workings of prejudice are not readily overcome. An entire generation of astronomers and physicists has been taught to accept – without any question – the idea of a Big Bang creation. It will not be easy for many scientists to admit that they have dismissed an alternative which should have been given further consideration. It becomes even more difficult when it is realized that this decision was based upon an obviously flawed fundamental scientific precept! Although it is almost inconceivable that present views of radiation propagation ever came to be established, this digression must be dealt with in the manner of genuine science – namely, through an honest and impartial analysis of all available evidence. In so doing, a truly amazing picture of the cosmos will be seen to unfold.

Nevertheless, it is also evident that the Steady-state cosmology, as presently understood, is very much in need of certain elaboration and modification. In particular, it must be clear that creation of elementary matter forms cannot possibly occur at the level of the hydrogen atom, which is now seen to represent a highly evolved state from the initial status of a lowly infinitesimal quantum, or "**A**." Also requiring clarification is the microwave background radiation, which is thought to argue in favor of an explosion from a superdense singularity. Similarly, many fundamental problems relating to the unique physical properties of matter remain unanswered.

The Construction of Matter

Without a doubt, a most pressing unsolved characteristic of matter is that of electricity, a strange influence which may be expressed as a form of either attraction or repulsion. While a force of extremely short range, it is an incredibly powerful phenomenon which completely dominates the nature of matter at close distance. But just what is electricity? Why do two "positive" or two "negative" charges repel each other; and why do two unlike charges attract?

A possible explanation suggests itself. Recalling the premise in which an inborn desire for fusion will induce a cloud of minute quanta to congregate about a material particle, it may well be asked if *halo rotation* might not afford a solution. Conceivably, we might theorize "**A**" as being created of opposite "charge" – or opposed directions of rotation – with electricity a consequence of highly concentrated and rotating energy fields coming into contact.

But how is it possible for a charged particle to always display one specific charge (or direction of rotation) relative to other such bodies – especially since it may approach at any one of an innumerable number of angles? Actually, this problem may arise since we are inclined to visualize rotation as something which transpires in but one direction at a time. By postulating simultaneous polar and equatorial rotations, it will be found that, no matter what angle two spherical bodies of *like* rotation come into contact, the motions of their respective surfaces will *oppose* or *clash*. Conversely, two bodies of *unlike* rotation will always *flow in harmony* with each other at the point where they touch. Hence, it is not difficult to infer a connection with the effects of repulsion and attraction, respectively. (The various types of particle spin theorized by physicists, in an attempt to explain certain phenomena, may be reconciled with such a concept on the grounds that they merely constitute *secondary* motions superimposed upon a much faster and more basic trait.)

It would therefore appear that we are justified in upholding the principle of opposed motion as a key to an understanding of this most fundamental characteristic of matter – a trait presumably initiated with the creation of "**A**" and perpetuated through the evolution of minute quanta to such ultimate particles as the electron and proton. On a philosophical plane, one might choose to view this electrical aspect of creation as a divided state somewhat analogous to our own sex-divided world – whereby an incentive exists for all "unbalanced" members to seek mutual balance and progress by means of union with oppositely "unbalanced" entities.

We may thus come to the conclusion that the mysterious phenomenon of electricity is but a unique manifestation of gravitation. In all probability, this force is simply a *highly condensed rotary flow of gravitational energy* – perhaps little wonder that our past failure to explain the nature of this communion should also coincide with the inability to define electricity.

In contemplating the evolution of matter from the lowly status of an "**A**," it must first be explained how these elementary forms are able to condense to a degree sufficient to produce such highly compacted and relatively massive structures as electrons and protons. Created initially in the great vacuum of outer space, in the voids between the clusters of galaxies, it is at once obvious that a sparse distribution of "**A**" must be transported to regions of much higher density – from whence the fusion process is enabled to function with vastly increased efficiency. The problem is thus seen to be one of transporting widely scattered quanta to a common meeting point.

In this regard, only the *repulsive* aspect of gravitation (along with radiation) surfaces as a viable mechanism. Converging from all directions in space toward the center of an expanding Creation Zone, between neighboring superclusters of galaxies, this steady streaming of energy facilitates a chain reaction effect. In time, a veritable deluge of elementary matter forms are induced to flow toward a central point – a region which eventually becomes of sufficient density as to ensure fusion into such mature structures as electrons, positrons and protons. Since the energy required to condense "**A**" is equal to that needed to push an equivalence of mass apart, it is tantamount to yet another demonstration of Newton's Third Law of Motion: "To every action there is always an equal and contrary reaction." Essentially, this exchange of energy accomplishes a twofold purpose, inasmuch as the *condensation of "**A**" is revealed to be achieved at the expense of producing an expanding universe!*

The next major problem involves the issue of particle size. How has nature contrived to impose the observed limitations in the sizes of such stable and "maximum" particles as electrons and positrons and the somewhat higher level of the proton? Invariably, rotation must hold the key. In compliance with established scientific law, which demands an increase of rotation with either an increase of mass or a reduction of diameter, it follows that a point will eventually be reached when a condensing body is revolving so fast that its outer surface begins to approach the speed of light. With gravitational communion thus negated, additional quanta will be rejected and the evolution of a "maximum" particle becomes self-terminating.

The fusion of elementary quanta into mature particles is linked to the ability to achieve gravitational communion. In order for any evolving structure to incorporate additional quanta it is vital that relative velocities be such as to allow *intimate association for a minimum interval of time.* Time must be given for a critical degree of communion to be established, including a possible exchange of the spiritual entities so involved. We can now see why radiation is unable to evolve to higher material structures. Due to its extreme motion through space, it is simply impossible to achieve the required time/proximity association with foreign quanta.

It will be inferred that the evolution of material particles may proceed as stated until the electron-positron level is reached, at which point rotation has grown sufficient to forbid further addition of quanta. Henceforth, evolution may continue only through entire electron/positron associations – a new and comparatively brief round of matter-building which terminates in the construction of the proton. Since, for a presumably similar reason, a proton also constitutes a "maximum" particle, it follows that continued evolution must involve a fusion of masses of the general order of protons to produce (in the interior of stars) the various elements. Complicated somewhat by a constant struggle of proton-neutron complexes to preserve unity, the factor of excessive mass/rotation may again act to terminate fusion.

As newly constructed electrons and positrons enter the dense and rapidly rotating nuclear region of a condensing primordial cloud, they will be seen to approach from a direction somewhat *perpendicular to the prevailing flow of energy.* Spiraling deeper into the core of the swirling cloud, a discrimination will soon arise with regard to the direction in which fusion is most likely to be facilitated. Indeed, there will be a preference for recent arrivals to be incorporated at the *leading edges* of developing proton-electron complexes – *producing protons only.* (Any antiprotons which are formed would be annihilated through subsequent encounters with overwhelming numbers of protons.)

Now it is obvious that contraction of a primordial cloud, with a steady acceleration in velocity of rotation, cannot continue without limit. For while an appreciable density is required at some stage of creation in order to expedite the evolution of material particles, the observed distribution of matter within the universe makes it apparent that this must be a very temporary state – invariably, *a state which precedes a mammoth explosion!*

The mechanism triggering this cataclysm must surely involve the tremendous rotary forces which exist deep within the cloud as it reaches

Cosmological Theories

an advanced stage of contraction, shortly following a large-scale production of protons. As the innermost regions approach the limiting speed of light, so a strong braking action will arise as material particles resist further acceleration in the vicinity of the supermassive black hole core of the cloud. Were it not for the force of material forms still streaming inward, the nucleus of the cloud would be induced to shatter at a relatively early stage. For some time this rising outward pressure may thus be contained until, at last, the vast bulk of all matter forms is blasted into space – *from whence the ejected particles will later condense into a supercluster of galaxies!*

Although on a scale which far surpasses that of a mere supernova eruption, widely differing circumstances will conspire to mask any outward signs of such a stupendous event. The exceedingly powerful gravitational field, combined with innumerable encounters that are highly conducive to fusion due to favorable relative motion, must facilitate the incorporation of all but an infinitesimal fragment of the cloud's total mass-energy into "maximum" particles. With gravity tending to prevent expelled radiation from fully achieving escape velocity, along with the presence of huge swarms of only partially evolved structures, it will be seen that an exceptionally strong incentive exists for free quanta to be promptly absorbed by these undeveloped material forms.

Vestiges of this cosmic matter-building process – implying an unending series of *Small Bang* explosions – are very likely to be found in the peculiar nature of *quasars*. Since it is most logical to infer that remnants of these superdense sites will persist, in one form or another, it is well to consider related features. The combination of extreme density and substantial mass, which is believed to constitute a unique characteristic of quasars, is in precise agreement with what might be presumed of a nuclear core left over from such an explosion. Indeed, the presence of a supermassive black hole at the center of a quasar is exactly what would be anticipated under the prescribed conditions.

Of all the "maximum" particles, it is the positron that is most suited to achieve escape into regions of future Creation Zones – providing subsequent generations with the necessary rotary impetus for the formation of protons. Due to a favorable mass-to-velocity ratio, they are small enough to be accelerated to escape velocity while yet of sufficient mass to overcome resistance through impacts with the surrounding quanta. Protons, on the other hand, cannot be imparted with the same velocity and must be contained by gravitation to form galaxies of stars. Attracted to protons by their compatible electric charges, almost all free electrons may be expected to be trapped into association – producing hydrogen atoms, the chief source of stellar fuel. (The production of primordial helium, beyond what is produced by stellar burning of hydrogen, is now seen to have an explanation in the concept of Small Bangs.)

The formation of neutrons may be looked upon as a process which occurs in the intense heat and density of stellar interiors. When a pair of protons are brought together with sufficient force to achieve a mutual penetration of halos, the two nuclei will remain in close association within the confines of a reconstituted overall halo which quickly assumes a characteristic proton rotation. By inferring the presence of an electron nucleus (produced out of the halos of the two protons' interacting energy flows), the stability of such a union may well be explicable in terms of two positive charges exchanging one negative charge. Thus, at any given moment, one of the proton nuclei will have temporary custody of the electron nucleus and may, in a sense, be considered to be the neutron.

Technically, the ability of an electron to neutralize the charge of a proton must embrace the principle of *interference,* whether it be a case of movement through an atmospheric halo or the interaction of nuclei at the center of such an enveloping cloud of energy. In the former instance, the disruptive influence of an oppositely rotating electron halo will serve to cancel outward effects of rotary motion. In the case of a neutron, this disturbance of the overall rotary flow is perpetrated from within. In short, a neutral particle may be defined as either a cancellation or absence of halo rotation. (The unstable nature of a free neutron is likely due to susceptibility of the electron nucleus to be ejected out of its eccentric orbit upon encountering an especially large quantum of energy at a strategic moment in time.)

The unstable family of mesons, with masses intermediate between protons and electrons, may be interpreted as segments of protons – bundles of electron-positron nuclei torn loose following violent collisions involving atomic nuclei. During such brief intervals (often less than one ten-billionth of a second!) both positive and negative characteristics may be exhibited, along with disturbing motions which would give rise to neutral properties. Tentative associations of this sort, including freak transient quarks and gluon-type peculiarities, are made possible by the inability of a system to change its mass-energy relationship in zero time. Even the rather elusive neutrino, a hypothetical particle of no apparent charge and virtually no mass, introduced by science to balance certain equations relating to nuclear events, may now receive interpretation as ejected infinitesimal quanta.

Cosmic Expansion and Energy Conservation

With the Steady-state principle of creation now revealed to be the prevailing cosmology, we are still left with somewhat of a loose end. In short, there is urgent need to reconcile energy conservation with the premise of Continuous Creation. Hitherto, most proponents of this model had simply glossed over such an issue as the actual size of our universe – some even to the point of assuming it to be infinite in physical dimensions. This is clearly a great mistake. Postulation of a model that is infinite in both time and creation poses an outright contradiction with respect to the "law of conservation of energy." How can we theorize an endless amount of new creation – even if the universe were assumed to be infinite in physical size? *How could any physical structure be more than infinite?*

The only solution to this paradox, of course, is to seek a means of reconciling new creation with the conservation of mass-energy. For regardless of consequences, there is but one way in which this dilemma may be avoided, and that is to infer *a disintegration factor precisely equal to the rate of creation!* For every atom or quantum of energy that is created, somehow and somewhere, an equivalence must undergo annihilation to balance the mass-energy books and thus preserve cosmic order. It merely remains to define the nature and circumstances of this truly momentous prediction.

As it turns out, the solution is already at hand in our revised views of radiation propagation. Invariably, reinstatement of the "law of addition of speeds" must raise the question as to what really does transpire at the edge of the universe, where radiation becomes totally trapped by the phenomenon of cosmic expansion. Without the slightest doubt, a most amazing situation must develop. As radiation proceeds inward, from an endless number of energy-emitting celestial objects, it encounters an equally endless stream of energy forms being driven outward by reason of an expanding universe. Since opposing velocities will be the same at the boundary of our physical universe, it follows that the end result must be a virtual entrapment of radiation which, in all essential respects, may be described as an encircling *"shell of energy."* In all probability, *the inferred disintegration occurs at the edge of our spherical universe, as receding stars and planetary systems alike strike this invisible barrier at the speed of light!*

So concentrated is this fateful energy field that it may, in fact, be likened to an *encircling black hole*. Quite invisible to the most powerful telescope, such a prodigious energy barrier must surely lie at the cosmic periphery where it acts to balance creation in a rather dramatic fashion. Subsequently, it may be deduced that many currently observed quasars and star systems exist only in the form of *ghostly images* – the real sources having long since vanished in response to a fundamental law requiring the conservation of mass-energy!

What are the chances of our Earth experiencing such a dramatic ending in the foreseeable future? Actually, the odds appear strongly in favor of just this happening long before life is extinguished as a consequence of stellar evolution. Regardless of the size of the cosmos, it may be computed that eight times as much matter resides in the outermost half of the universe (from the point where recession is 50% that of light) as in the remaining portion – all of which must undergo annihilation in one cycle of Hubble's constant. By the laws of probability, it might be argued, we are presently long overdue and living on borrowed time!

True though this may be, the immediate situation is not nearly as frightening as it might first seem. Inside the very structure of the universe, a human lifetime represents the smallest fragment of time in the overall sequence of events. Even conceding that we may now occupy a position very close to the edge, the prospects of us reaching this fateful boundary are extremely slim in terms of a decade or a century. Should we substitute "millions of years" as the time frame it would, conceivably, be quite another matter.

4

New Horizons in Cosmology

The "Small Bang" Scenario – Quasars as Celestial Beacons – D/R Arcing of Radiation – Redshifts as Distance Indicators – The Nearby Cosmic Edge – The Enigma of Galactic Redshifts – The Microwave Background

The value of any scientific theory lies in its ability to make predictions and withstand tests. Unless it is susceptible to analysis and criticism, it must remain an idea rooted solely in conjecture. Most assuredly, the premise of quasars as Small Bang vestiges is a verifiable concept which invites urgent investigation.

The "Small Bang" Scenario

In portraying quasars as vestiges of Small Bang creation sites, it is well to consider the factor of intrinsic luminosity. Conceding the strong possibility that mergers of colliding galaxies could produce luminosities which mimic quasars, it is perhaps in order to adopt a minimum absolute magnitude of -23.0 as a means of identification. It is also essential to realize that quasars tend to be far more luminous in their youth, when there is much more gas and dust available to fuel their supermassive black hole dynamos. The actual length of time in which a quasar is able to shine as such, before evolving into a giant compact galaxy, is subject to a degree of speculation. (Best current estimates would seem to range between several hundred million and several billion years.) Almost certainly, the massive elliptical galaxy known as M87, in the constellation of Virgo, was once a quasar parent of our own Milky Way system in the distant past.

79

This quasar interpretation of Small Bang vestiges can be put to an interesting test. Having some knowledge of the average separations, between the centers of neighboring superclusters, it becomes possible to calculate the numbers of quasars compatible with this prescribed model. Furthermore, it is feasible to compute distribution in terms of Z redshift. However, before attempting to reduce the abundance of quasars to formula it is desirable to contemplate just what effect disintegration at the edge of the universe will have upon the maximum redshift which a quasar may exhibit. For quite regardless of where a quasar or galaxy may form, within our finite expanding universe, there must be an upper limit as to how long it can continue to exist before reaching the velocity of light – relative to the cosmic center – and suffering extinction. Subject to certain subtleties due to proximity to the cosmic edge, this cutoff of Z redshift approximates the maximum interval of time that light can remain in transit – *without either the radiation or ourselves going over the edge.*

Assuming a separation of roughly a quarter-of-a-billion or so light-years as a typical initial distance between a recent Small Bang site and the closest supercluster, it is possible to assess certain implications of this figure as it applies to the total number of shells or cycles of Hubble's constant likely to characterize our physical universe. In effect, the universe may be visualized as a series of concentric shells expanding outward from a common center, with each successive shell double that of all the inner shells. In billions of light-years (BLY) the radius of each successive shell, moving inward from the outer edge, is as follows: 1) 9 BLY; 2) 4.5 BLY; 3) 2.25 BLY; 4) 1.125 BLY; 5) 0.5625 BLY; and 6) 0.28125 BLY. Upon this basis there will be seen to be some six cycles of Hubble's constant and a minimum average separation between recent Small Bang zones and previous superdense sites (by now superclusters of galaxies) of about 0.28125 BLY. (See Figure 5.)

Thus, in a period of 12 billion years, a centrally-located Small Bang creation site will have moved to a distance of about 0.5625 BLY and will be receding at 3.125% of the velocity of light. In two cycles of Hubble's constant, it will have doubled both speed and distance, reaching a point some 1.125 BLY from the center of the universe and possessing a speed 6.25% that of light. After three cycles, there will be a distance of 2.25 BLY and a velocity 12.5% that of light. With four cycles, distance will be increased to 4.5 BLY and speed of recession will now be 25% that of light. Five cycles will result in a distance of 9 BLY and a velocity of 50%. Upon achieving six cycles (equivalent to $Z = 4.0$ and a transit time of 72 billion years), this same region of space will have attained the velocity of light, relative to the center of the universe, and will have moved some 18 BLY to the cosmic edge where sudden extinction awaits. (It is to be mentioned that redshifts, well in excess of $Z = 4.0$, are to be expected by an observer situated in close proximity to the nearby edge, due to the almost virtual entrapment of radiation at the cosmic periphery. Also, the very premise of a general bending of light must imply enhanced transit times – beyond $Z = 4.0$ – for certain remote sources located at the other side of the universe.)

Upon such a theoretical basis the gradual increase in numbers of quasars at low redshift, along with a dramatic and steady rise up to a point well below $Z = 3.0$, is exactly the combination of curve (number observable) and slope (number if there were no extinction) required to fit observation. (See Figure 6.) Moreover, the quantities implied at various redshifts appear to be in excellent agreement with evidence. (Predictably, the frequency of Small Bangs – within the expanse of our universe – is likely to be about one in every 10^5 years.)

A departure from the general slope, at higher redshifts, has long been noticed by astronomers who have chosen to interpret the entire curve in terms of subsequent condensations from a solitary Big Bang creation. Accordingly, it was assumed that the birth of quasars began slowly almost two billion years after the initial explosion – with formation proceeding at an accelerated rate until finally stabilizing some billions of years later. (The problem of explaining the equally pronounced curve below $Z = 1.0$, in which the population of quasars is seen to dwindle to zero at a distance of a little over half-a-billion light-years, remained a rather good mystery. Why should there be no quasars in our Local Supercluster, while they are abundant only a billion or so light-years away and with relatively minor difference in time of light transit?) Not only does the concept of Small Bang creation vestiges readily account for the observed curve at low redshifts, but it also affords an excellent explanation for the noted departure at high redshifts. Indeed, this last curve can now receive a very different interpretation once a proper assessment has been made with respect to either a literal disintegration or near entrapment of radiation at the boundary of our universe.

Incidentally, impressive evidence, confirming a Small Bang/Steady-state cosmology, may be adduced from a symmetrical "honeycomb" pattern which has been revealed in the distribution of galaxy superclusters.* Quite

---

* "Large-scale distribution of galaxies at the Galactic poles," *Nature*, February 22, 1990.

inexplicable in terms of any Big Bang creation, it is observed that remarkably periodic separations characterize the spatial arrangement of superclusters. This highly unexpected homogeneity is precisely what must be inferred upon the basis of our newly prescribed model! Not only is just such a pattern mandated, but there is also seen to be excellent agreement with regard to theoretical separations and the actual distance intervals measured in this deep redshift survey.

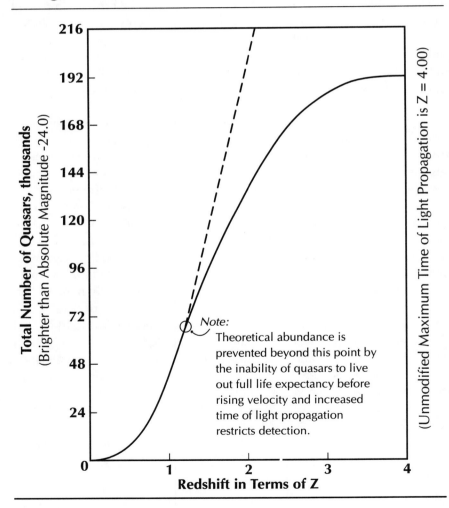

Figure 6. **Observable Quasars in Terms of Z Redshift**

Note: Theoretical abundance is prevented beyond this point by the inability of quasars to live out full life expectancy before rising velocity and increased time of light propagation restricts detection.

Now the very geometry of a basic Steady-state cosmology, which includes Small Bang creation sites, must mean that quasars (of the same age) likely possess typical separations with regard to spatial distribution. Totally against the dictates of a solitary superdense origin, if such a preferred separation factor can be substantiated it will afford even more graphic proof of an Eternal Universe.

Unfortunately, any separation pattern will tend to be masked by a number of formidable factors. A variation in the sizes of Small Bang creation sites, for instance, must have an obvious effect. So will a difference in the ages at which quasars are viewed. Of concern at close range, any age differential will become of major importance when more extreme distances are involved, as it must lead to increased movement away from the point of their birth. Also compounding the problem is the deduced arcing of extragalactic radiation, along with a pronounced distortion of redshifts which is to be found in the general direction of the nearby cosmic edge. Nevertheless, in spite of these obstacles, there is hope that it will eventually become possible to confirm the prediction of a common factor of separation between quasars of similar age. It may well be proven that mysterious cases of quasar redshift "clumping" – previously looked upon as statistical curiosities – have a perfectly logical solution within the framework of a Steady-state/Small Bang model.

Quasars as Celestial Beacons

Considering the fact that we see quasars as they were billions of years ago, we also see them as appearing in positions distorted by the curvature of their light as it is influenced by the expansion of the universe. By way of recapitulation, this displacement of images (or "cosmic arcing") is determined by three factors:

1) The *time* of light transit, as implied by distance. (A longer transit time must result in increased curvature.)

2) The *position angle* of a quasar, relative to our motion toward the edge of the universe. (The greater this angle, the greater the bending of light.)

3) The *proximity* of an observed quasar to the outer edge, in which the curvature of radiation is seen to be enhanced with increased distance from the center of the universe.

In every instance, displacement will be one *in which light is bent toward the cosmic center.* (See Figure 5.) The inevitable question now arises, of course, as to whether observation can substantiate such a prediction.

Should we happen to be situated in a relatively central portion of the universe (with little motion toward the edge) we might expect to see no appreciable clumping of quasars, which would appear evenly distributed in all directions. Upon the far more logical basis of an eccentric location, and assuming the validity of our new cosmological model and revised views of light propagation, one should anticipate an unhomogeneous pattern with regard to the distribution of quasars. The standard Big Bang cosmology, on the other hand, offers two choices. With a somewhat improbable central location, a uniform distribution is again indicated. In contrast, an eccentric position would require the complete absence of all high redshift objects in one very specific direction – namely, toward the nearest edge of the universe. (By the "laws of probability," we should be situated at least 14 BLY from the cosmic center and no further than 4 BLY from the nearby edge, since slightly more than half of all the matter in the universe is located in this latter zone.)

To date, a weakness of all quasar surveys may be ascribed to the relatively small number cataloged and to the somewhat less-than-ideal homogeneity with regard to the manner in which some quasars have been found. Nevertheless, while such unfortunate bias has yet to be negated, and is likely to persist in one form or another for years to come, it is extremely doubtful if the overall scenario will be unduly affected by current limitations. What is presently available is really tantamount to a typical sample poll, from which a great deal of useful information may be obtained concerning the structure of our dynamic universe. Indeed, rich fields of quasars will be seen to exist in spite of preferential treatment; in fact, they were the very reason for intensive study! Likewise, regions exhibiting few quasars have been unjustly avoided simply because they had already proven to be unfruitful fields.

Upon detailed analysis of a comprehensive quasar survey, published by the *European Southern Observatory,** a most instructive pattern is noted. Containing a total of 2,720 members (of at least absolute magnitude -23.0), it is at once evident that the vast bulk of quasars reside in two major

---

* M.-P. Veron-Cetty and P. Veron, *A Catalogue of Quasars and Active Nuclei (2nd Edition),* Scientific Report No. 4, April 1985.

clumpings. Not only are these two concentrations situated in exactly opposite directions of the sky, but a conspicuous shortage is observed at right angles to such aggregations – posing a distribution which is totally against the dictates of the popular Big Bang cosmology! (See Figure 7 and Figure 8.) Moreover, when only the most luminous of these quasars are considered as a group, it is revealed that an overwhelming proportion of high redshift/high luminosity quasars are centered about the constellation of Sculptor in the Southern Hemisphere – a region which is also seen to be essentially devoid of high luminosity quasars of low redshift! In contrast, the opposite hemisphere is characterized by a mixture of high redshift and low redshift quasars.

Although it might be thought that the center of the universe must lie in the midst of the high redshift clumping, this is most certainly not the case. Upon the basis of a Small Bang/Steady-state cosmology, and an eccentric cosmic location, the central regions of our universe should display a wide variety of redshifts. On the other hand, the nearby edge is bound to be deficient in brilliant low redshift quasars – simply because there is less room for any new and highly luminous quasars to be formed! Furthermore, a concentration of high redshift objects – toward the immediate edge – is to be expected by reason of the virtual entrapment of radiation, which allows us to view multiple generations of quasars with greatly extended light transit times. Accordingly, it will be concluded that the cosmic edge is probably located near the coordinates of right ascension 1 hour and -30 degrees declination (R.A. 1^h, -30° Dec.), very possibly within the constellation of Sculptor.

Quite inexplicable in terms of any Big Bang interpretation, the mystery of a dichotomous quasar distribution is therefore seen to have a perfectly logical solution. It is clearly much more than a mere coincidence that the congested central core of the universe is also accompanied by a pronounced right angle shortage – a phenomenon which is invariably produced by the prescribed bending of extragalactic light. The prodigious edge group clumping, with its preponderance of high redshift quasars, has a ready explanation upon the basis of our eccentric cosmic location and a long overdue revision of our view of radiation propagation. Instead of visualizing such highly redshifted radiation as having come from the outskirts of the universe many billions of light-years away, it is really our own Milky Way which has done most of the traveling! Moving outward more than half the radius of the universe since the formation of our galactic system, we

Figure 7. Northern Hemisphere Quasars

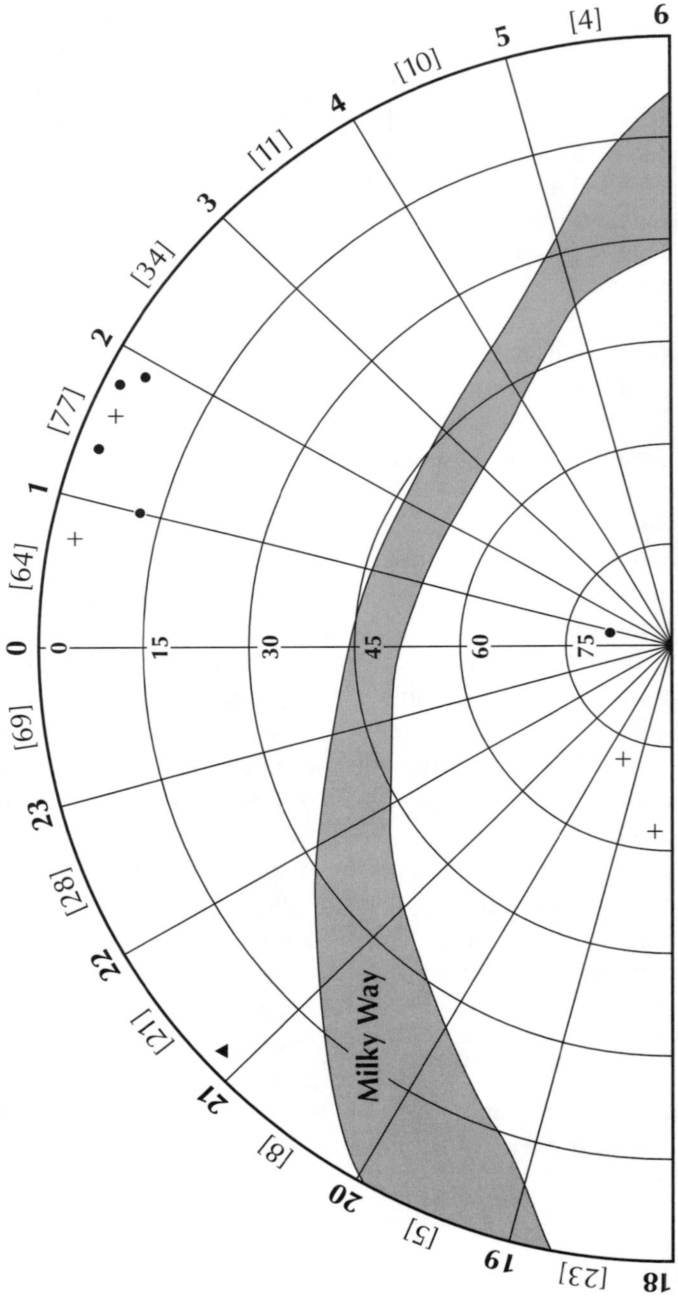

New Horizons in Cosmology

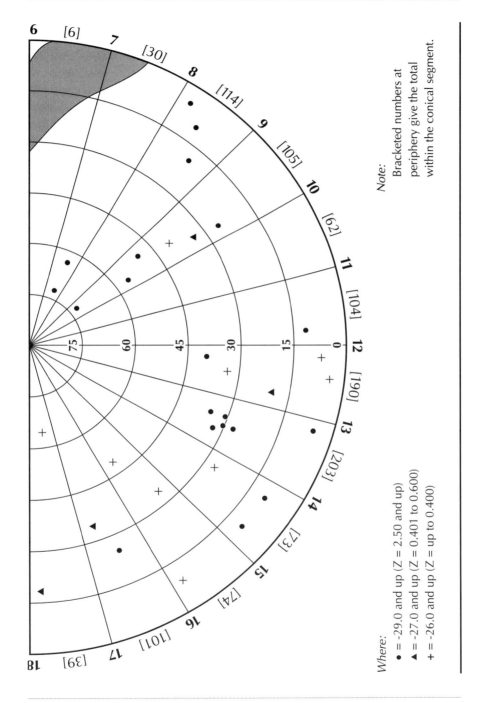

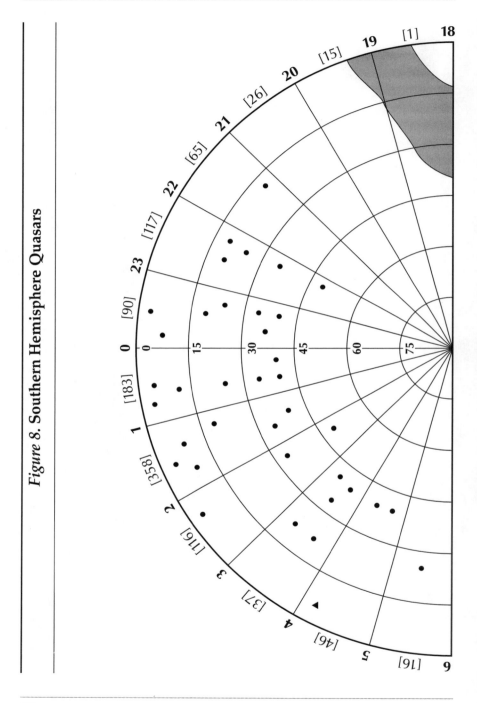

Figure 8. Southern Hemisphere Quasars

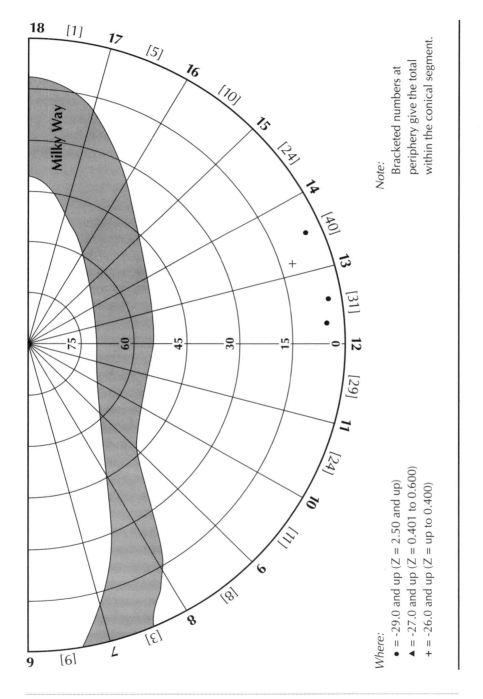

are only now encountering the light of long extinct quasars which has literally hung suspended – *almost in a state of total entrapment* – close to the cosmic edge for truly enormous periods of time. When we observe high redshift quasars, with light transit times of tens of billions of years, we are actually viewing very ancient images which might well be described as *celestial fossils* – the real objects having ceased to exist eons before our Earth was born. It may be presumed, in fact, that – in the direction of the nearby edge – all quasars (and most galaxies) are nothing more than ghostly apparitions of long extinct phenomena!

With quasar luminosity acknowledged to be a function of age, it is expedient to assess the distribution of quasar luminosity versus their redshifts over the expanse of the entire sky. In searching out high luminosity quasars of all visual magnitudes, and adopting a lower limit of absolute magnitude -29.0 and a redshift minimum of $Z = 2.5$, we find a total of 65 qualifying quasars in the aforementioned *European Southern Observatory* catalog. Upon plotting these objects on the two hemisphere charts (Figure 7 and Figure 8), it is at once evident that a preponderance (with a ratio of 41-to-24) is centered upon the coordinates of the inferred nearby cosmic edge. In point of fact, almost 2/3 of these exceptional objects are situated in an area comprising only 1/3 of the heavens! This translates to a ratio of more than 5-to-1 in favor of an unhomogeneous distribution. (When magnitudes of at least -30.0 are substituted, the ratio is increased to 6.5-to-1.)

Of equal significance is a pronounced shortage of low redshift quasars of high luminosity residing in the vicinity of this same nearby cosmic edge, *where they are quite conspicuous by their absence!* Most abundant elsewhere, they are almost totally absent in this one particular direction. (See Figure 7 and Figure 8.) Needless to say, an explanation is demanded for yet another strange anomaly. Impossible to answer in terms of any Big Bang cosmology, this peculiar shortage will now be conceded to arise by reason of our highly eccentric cosmic location, as reduced volume of space acts to impose a rather drastic restriction as to where (and in what direction) luminous quasars of recent vintage may be formed. In this one preferred direction of the sky, it must surely follow that the last light of many neighboring quasars has already passed us en route to more central regions of our universe.

Valuable insight into this most biased distribution may be found through analysis of the quasar sample following its division into a succession of redshift bins. Expressed in multiples of $Z = 0.200$, and considering an edge group member to be one residing in this same 1/3 portion of the total sky, we are enabled to compile the data conveyed in Figure 9 and Figure 10. A truly amazing picture emerges. With virtually a zero difference at $Z = 0.200$, substantial excess luminosity (away from the edge) is quickly encountered and observed to peak in the vicinity of $Z = 0.400$ – *from whence the luminosity differential is seen to proceed in the opposite direction!* Effecting a complete reversal between $Z = 1.80$ and $Z = 2.00$, a progressive surplus of intrinsic luminosity (in the direction of the edge) will be noted to characterize the remainder of the higher redshift bins. Nor is this variation of a minor nature. From $Z = 0.400$ to the limit of detection, the

Figure 9. Quasar Luminosity by Redshift/Direction

Redshift Range, $Z =$	Magnitude Difference		Quantity	
	Average Magnitude (diff. per bin)	Average Magnitude (total diff.)	Edge	Balance
up to 0.200	0.0061[a]	0.0061[a]	22	60
0.200 to 0.400	0.5572[a]	0.3232[a]	67	98
0.400 to 0.600	0.2007[a]	0.2458[a]	67	103
0.600 to 0.800	0.1688[a]	0.2076[a]	66	102
0.800 to 1.00	0.3000[a]	0.1788[a]	71	90
1.00 to 1.20	0.1200[a]	0.1715[a]	60	94
1.20 to 1.40	0.1101[a]	0.1959[a]	61	109
1.40 to 1.60	0.0688[a]	0.0887[a]	96	100
1.60 to 1.80	0.2517[a]	0.0689[a]	91	102
1.80 to 2.00	0.3061[b]	0.0645[b]	140	145
2.00 to 2.20	0.2824[b]	0.2275[b]	212	144
2.20 to 2.40	0.1348[b]	0.3061[b]	206	106
2.40 to 2.60	0.4996[b]	0.3108[b]	50	59
2.60 to 2.80	0.8436[b]	0.3316[b]	38	36
2.80 to 3.00	0.5337[b]	0.3482[b]	29	22
over 3.00	0.8764[b]	0.3696[b]	39	35
			1,315	1,405

[a] Brighter away from edge.
[b] Brighter toward edge.

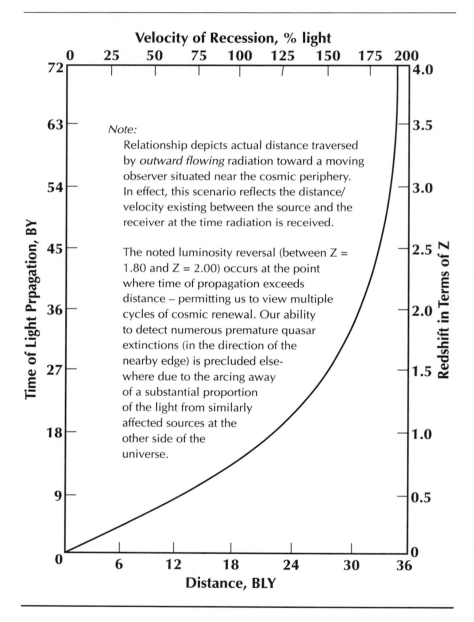

Figure 10. **Redshift/Time/Velocity/Distance Scale (Eccentric View)**

combined bin differential average translates to a brightness factor of some 360% – a factor which is direction oriented! Even when modified appreciably, through considering only the total average luminosity of all quasars within this same spread of redshifts, we are left with an overall luminosity surplus that is 174% higher in the direction of the nearby edge. Most assuredly, this striking pattern is not at all compatible with any explosive Big Bang origin.

In the process of explaining this remarkable observation, it will be seen that we are now in a position to verify a proffered solution for a major weakness inherent in previous versions of Steady-state cosmology – namely, the lack of conservation of mass-energy. Indeed, very graphic confirmation of this momentous prediction (involving an annihilation factor at the cosmic periphery) is already at hand in the evidence afforded by Figure 9, in which there is clearly a tendency of intrinsic luminosities and extreme redshifts to be noticeably higher in the direction of the nearby edge. Inasmuch as the energy output of a quasar is bound to diminish with age (typically, a hundredfold drop may be inferred in the span of about 10 billion years), as it slowly evolves into a giant galaxy with a supermassive black hole core, it follows that *higher luminosity is also a definite indication of youth!* But why should they be younger? This is surely the question which begs an answer. The obvious conclusion, of course, is that something must be preventing normal aging of quasars in this one region of the sky. Invariably, *it is a selective effect of premature disintegration at the boundary of our physical universe* – from whence an equivalence of mass-energy is subsequently reincarnated as infinitesimal quanta in the great voids of space.

In order to acquire deeper insight into certain aspects of our eccentric cosmic location, it is desirable to contemplate the path of radiation as it streaks toward us from the opposite side of the universe. Fighting a strong outward flowing stream of quanta during initial stages of its journey, it will spend tens of billions of years merely to reach the cosmic center. Thereafter, it will experience a "tailwind" and is enabled to cover the remaining 18 BLY in a period of barely more than 12 billion years. (See Figure 5.) Thus a study of extreme quasar redshifts – in the precise direction of the cosmic center – is capable of providing knowledge as to our proximity to the nearby edge of the universe, since it will reflect the time (hence distance) that radiation has been in transit to us. Tentative study of the evidence would seem to place us in an extremely eccentric position – perhaps, no more than scant millions of light-years from the fateful cosmic boundary that is believed to lie in the direction of the constellation of Sculptor.

It will also be acknowledged that the redshift/distance curve of Figure 4 really depicts the scenario to be expected from a somewhat centrally located observer. A more realistic view, looking from a highly eccentric location toward the center of the universe, may be found in Figure 10. The aforementioned luminosity reversal, between Z = 1.80 and Z = 2.00, will now be seen to represent the point where *time of propagation overtakes distance* – beyond which a rapid increase in transit time is effected with only a nominal increase of distance. In essence, at time of radiation emission, many of the more centrally situated quasars were really further away (hence more luminous) than had been believed upon the basis of traditional theory. (This need to revise upward current distance estimates of many centrally appearing quasars will not detract unduly from the inferred picture of premature annihilation at the cosmic edge, since many of these images have been strongly redshift-enhanced by reason of cosmic arcing.)

D/R Arcing of Radiation

The very premise of distance/repulsion (D/R) arcing of radiation affords a variety of observational tests. In addition to the large-scale phenomenon of cosmic arcing, in which distant quasars are revealed to have a most peculiar dichotomous distribution, we are led to infer yet another and more localized instance which might be termed "galactic arcing."

Impressive statistics have been advanced which clearly indicate that a disproportionately large number of quasars are to be found in close angular proximity to relatively nearby massive galaxies. Quite inexplicable upon the basis of traditional views, such an association suggests a physical connection which would be incompatible with the principle of redshifts serving as reliable distance indicators. It therefore remains to be explained how this anomaly can be resolved in the face of strong conflicting evidence that at least some quasars are at distances which are essentially commensurate with their redshifts.

Notwithstanding that the proposed cosmic aberration factor was able to explain an opposed clumping of quasar images, it is evident that this general arcing will be insufficient to account for such a quasar/galaxy anomaly. Yet another subtle mechanism must somehow be involved – *one which would permit these seemingly associated galaxies to focus incoming quasar radiation toward such centers of mass!* In all probability, it is the proffered D/R factor which is responsible for creating a scenario of high redshift quasars appearing to be connected with low redshift galaxies.

Just as a steady outward flowing of D/R quanta must produce an expanding universe and a large-scale displacement of images toward the cosmic center, so a massive intervening galaxy is able to induce a small-scale curvature in the path of radiation approaching from a more remote source – in this instance, curvature *away* from the less distant concentration of mass. However, the very action of "repulsive" quanta pushing against such approaching radiation must ensure that whatever does reach us will do so from a somewhat different angle – namely, *from a direction considerably closer to the intervening galaxy!* (See Figure 11.) Subsequently, the apparent overabundance of quasars – in angular proximity to major galaxies – is really an illusion produced by D/R arcing, and this hitherto unrecognized aspect of gravitation is thus seen responsible for another example of celestial deception.

Strong statistical evidence exists in support of the concept of "galactic arcing," by which the aforementioned D/R factor is seen to be responsible for producing the illusion of a disproportionate number of quasars seemingly associated with massive galaxies. In order to achieve this clumping, in the angular distribution of quasars, it is necessary for these images to have been displaced from regions which are adjacent to an intervening galactic body. Remarkable confirmation of this prediction may be found in the research of several prominent astronomers,* in which a dearth of quasars is indeed shown to be a feature characterizing the outlying regions surrounding just those massive galactic systems in question.

Among the many astronomical anomalies which have been uncovered are instances where luminous bridges appear to physically connect two objects of widely differing redshift, thereby negating the worth of extragalactic redshifts in formulating any distance/velocity scale. There is now every reason to believe that a solution to such enigmas may be found in

---

* Among the more astute scientists to pioneer this important line of research are Dr. Halton C. Arp and Dr. Geoffrey Burbidge, whose published works have inspired much controversy and cast doubt as to the credibility of many accepted views. An informative account of such statistics is summarized in an article by David Cherry, entitled "Redshifts and the Spirit of Scientific Inquiry" (*21st Century Science & Technology*, May-June, 1989.)

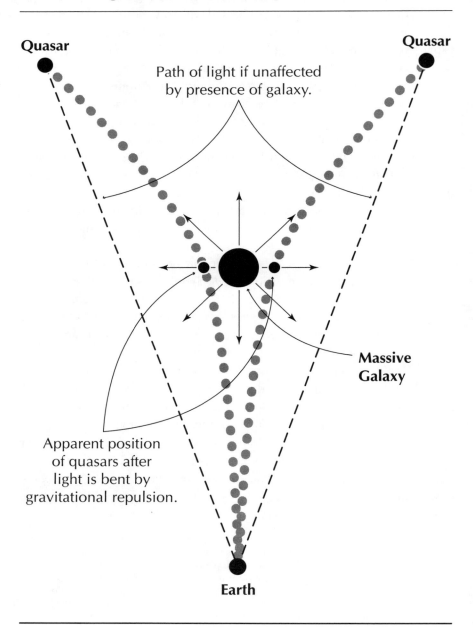

Figure 11. **Galactic Arcing of Radiation**

the concept of *cosmic arcing*, as imposed by the D/R factor responsible for our expanding universe.

Perhaps the most notable example of a luminous bridge involves a galaxy known as NGC 4319 (Z = 1,700 km/s) and the quasar Markarian 205 (Z = 21,000 km/s). Located at the coordinates of R.A. $12^h\ 21^m$, and +75° 45' Dec., the quasar is situated almost due south of the galaxy. Close examination of this pair reveals a narrow but quite distinct connecting bridge. At first glance, this would appear to afford irrefutable proof of physical proximity, being located at the edge of the halo of gas and stars surrounding the galaxy's nucleus.

It is highly probable that this illuminated bridge is a product of cosmic arcing. In the course of having its radiation arced toward the center of the universe, the distant quasar's light will be seen to have passed behind the nucleus before emerging in the vicinity of the galactic halo. During this grazing passage any intervening matter will be light-enhanced, in the form of a narrow illuminated beam or path, by the relatively intense radiation of the quasar. It may well be much more than a coincidence that this bridge points precisely toward the cosmic center!

Incidentally, this general deflection of images — toward the center of the universe — may explain a curious distribution of quasars about the galaxy M82. (The coordinates of this system are R.A. $9^h\ 51^m$, +69° 55' Dec.) No less than four quasars appear to lie in a somewhat narrow cone stretching to the southeast of the galaxy — a direction which just happens to coincide exactly with a line extending toward the cosmic center! It is difficult to refute the inference that we are looking at an example of cosmic arcing reinforced by a more localized galactic focusing of images. Needless to say, such instances serve to emphasize the importance of conducting an unbiased search for other examples which would further confirm this most logical interpretation of observation.

Nor is this general cosmic arcing of images confined to distant quasars. On the contrary, should our Milky Way be situated in close proximity to the cosmic edge — as all the evidence strongly suggests — we would then be in position to observe a most peculiar clumping of galaxies. In effect, the outflowing stream of D/R quanta will cause the images of many right angle/edge group exterior systems to be transposed considerably toward the center of the universe — producing the mirage of elongated sheets of galaxies stretching far across the heavens. It is to be noted that such a striking celestial phenomenon has already been discovered by astronomers, who have chosen to refer to this bizarre distribution as the "Great

Wall." (Incidentally, the so-called "Great Attractor" is likely a companion effect of cosmic arcing. This curious galactic streaming has led some astronomers to postulate an unseen attractive force some hundreds of millions of light-years away. However, it will now be seen that an appropriate title might well be the "Great Illusion.")

Adding to the confusion is data which would seem to establish a rather peculiar pattern in galaxy redshifts. To the astonishment of many scientists, it has been disclosed that the redshifts of galaxy cluster members tend to congregate about certain discrete values – implying that an unknown subtlety is involved in the creation of such redshifts.* Understandably, this revelation has served to add impetus to the controversy regarding the credibility of redshifts as distance indicators.

Although much remains to be studied in depth, it now appears that this phenomenon is rooted in one aspect of D/R arcing. In the process of interacting with radiation the D/R quanta must imprint a time delay which is independent of actual curvature. In effect, when radiation approaches our Milky Way Galaxy it encounters a specific strength – related to the mass of our system – of "repulsive" quanta, which acts to increase the time of transit. (The greater the mass of a receiving galaxy the greater must be this retardation factor that is imposed upon incoming radiation, as absorption and emission of quanta cannot occur in zero time.) Thus, it will be seen that our own Milky Way is responsible for literally imprinting a specific degree of time delay (or redshift) upon all extragalactic radiation.

This interpretation can be put to an interesting test. Upon the basis of theory, radiation received from low-mass galaxies should be somewhat more redshifted than from equidistant high-mass systems – simply because the strength of "repulsive" quanta emitted from the more massive galaxies will be greater. In turn, this stronger flow will tend to facilitate the propagation of outward-bound radiation in the face of an opposing flow streaming from our own system. Analysis of the redshifts of nearby galaxy clusters affords confirmation of this prediction. In the M31 (Local Group) and M81 clusters, both of which have been subjected to close scrutiny, all 21 major companions (of relatively low mass) have been found to possess higher redshifts than their more dominant cluster constituents! The chance of this being an accidental occurrence is so incredibly slim that it must lend strong support to the principle of a D/R factor.

---

* W.G. Tifft and W.J. Cooke, "Quantized Galaxy Redshifts," *Sky & Telescope*, January, 1987.

There is reason to suspect that a noted "quantization" of galactic redshifts among members of such galaxy clusters as Abell 262 and the Coma Cluster, for example, is similarly rooted in encounters with D/R quanta. To the puzzlement of observers, it has been discovered that different types of galaxies (lying at the same distance) actually display different redshifts. Those high luminosity galaxies of greater mass tend to possess lower redshifts than their less brilliant and less massive neighbors. It may now be suggested that this peculiarity affords yet another example of the D/R factor at work. In essence, the overall flux of "repulsive" quanta (flowing from galaxies) will impart very distinctive patterns with regard to time of light propagation – hence, to redshift. (The magnitude of this effect would imply that redshifts beyond $Z = 4.00$ may not be as uncommon as might otherwise be thought, since a lengthy transit of radiation is bound to involve numerous encounters with discrete concentrations of D/R quanta – leading to an appreciable increase in time of propagation.)

One further cause of seemingly "quantized" redshifts is bound to be related to geometrical effects inherent in the "New Cosmology." By reason of this hitherto unrecognized repulsive factor many galaxies will be induced to exhibit hollow globular structuring within the confines of an overall supercluster, as the D/R force acts to push such members to separations consistent with age and mass. Viewed from afar, a "bubble-like" sphere may be expected to present three distinct instances of redshift clumping. In addition to the foreground systems attempting to approach us due to repulsion from within the cluster, so an opposing background swarm will be receding at higher speed (with respect to ourselves) from the opposite side of such an expanding "bubble" configuration. Another above average concentration will take the form of a dichotomous right-angle distribution that is seen to be intermediate in distance between these two extremes. Accordingly, a majority of galaxies will portray characteristics of a "quantized" nature, as the factor of gravitational repulsion is permitted rather graphic expression over distances measured in tens of millions of light-years.

Redshifts as Distance Indicators

To assess the worth of extragalactic redshifts as a viable means of determining distance, it is helpful to review certain fundamentals implicit in the revised concept of radiation propagation. First and foremost, a redshift is a true measurement of the elapsed *time* between emission and eventual detection – quite regardless of direction and any conceivable form of intervention experienced by radiation while it is en route. In order to convert time into distance it is essential that the factor of cosmic expansion be taken into consideration, as radiation is caused to propagate at the velocity of light *relative to the moving field in which it is embedded!* With both direction and eccentricity of our cosmic location regulating the approach of radiation, we are led to contemplate such issues as luminosity and angular displacement in terms of redshift.

The luminosity of an object is a measure of the number of photons received in unit time. Assuming no intervention, the traditional inverse-square law will apply to the actual distance between source and observer's "location" at the time of light emission. In this instance "location" is defined as that region of space which, in full compliance with the dictates of cosmic expansion, would intercept such radiation in the interval of time reflected by redshift – regardless of whether this region of space is occupied by receiver's world at the moment of emission. Restated slightly, with "location" situated in the direction of the cosmic center, distance will depict the time required for expansion to move it to the position of a source residing toward the edge. With regard to a source located at the other side of the universe, distance will express the propagation of radiation through changing flows of quanta. (After passing through the cosmic center a "headwind" will be replaced by a "tailwind.")

The angular displacement of an object is analogous to luminosity, also being commensurate with the inverse-square law applied over the distance between source and receiver's "location" – within a steadily expanding universe – at time of light emission. Accordingly, angular size will not change with an observer's proximity to the cosmic edge, and will therefore always exhibit the same relationship to distance as luminosity.

Extragalactic redshifts, however, while affording an accurate view of elapsed time, are seen to be very susceptible to misinterpretation when one attempts to convert measurements to velocity of recession and distance – leading to numerous instances of confusion and an outright contradiction with regard to a variety of celestial associations.

Compounding any redshift/distance relationship is yet another hitherto unrecognized ingredient. Contrary to abundant evidence, and ignoring the research of certain astute scientists, there has been a general reluctance to accept the idea that interstellar hydrogen gas could act to enhance redshifts.* The fact that radiation is known to propagate faster in a vacuum than through other mediums, such as air or water, is conveniently glossed over by a scientific community quite unwilling to jeopardize traditional Big Bang redshift/recession statistics. In the last analysis, the only issue at stake is really one of assessing the magnitude of this additional "time delay" factor.

Although it might be tempting to theorize that the redshifts of all distant galaxies and quasars could be explained on grounds of interaction of radiation with extragalactic hydrogen, there is good reason to believe that any enhancement must be rather limited – if not in specific and localized instance, then at least in overall extent. Quite simply, the distribution of such hydrogen gas is bound to be far from homogeneous. Indeed, upon the basis of the "New Cosmology," it is to be inferred that the vast bulk of this gas will be confined to relatively small pockets – largely in the vicinity of recent Small Bang creation sites. Thus, it may be deduced that a modest degree of such radiation delay undoubtedly exists, where it must add to the redshifts of distant celestial phenomena. However, it is likely that instances of extreme redshift enhancement are confined to infrequent occasions where radiation has passed through exceptionally dense concentrations of extragalactic hydrogen gas. (Conceivably, this could afford a possible explanation for certain otherwise enigmatic astronomical anomalies.)

The Nearby Cosmic Edge

In a classic instance of serendipity, the existence of certain anomalous galactic redshifts presents a much needed opportunity to more accurately determine our proximity to the nearby edge of the universe. In effect, a relatively small radial separation – between two gravitationally interacting systems residing close to the cosmic edge – is capable of displaying highly discordant redshifts. Hence, calculation of true separation, as a function of respective light transit times, offers a means of deducing our present location with respect to a fateful cosmic boundary.

---

* See "A New Non-Doppler Redshift," by physicist Paul Marmet (*Physics Essays*, Vol. 1, pps. 24-32, 1988). This principle is also expounded in another paper by the same author, entitled "Red Shift of Spectral Lines in the Sun's Chromosphere" (*IEEE Transactions on Plasma Science*, Vol. 17, No. 2, April 1989).

One promising galactic association is that of NGC 7603, a large Seyfert galaxy with a smaller companion system evidently attached by a curving filament of luminous stars and gas. The redshift of the larger galaxy is indicative of a traditional recession velocity of some 8,700 km/s, while that of the smaller galaxy is 17,000 km/s. If we were to tentatively place the Milky Way at a current distance of some 3 million light-years (MLY) from the cosmic edge, the effective approach speed of any incoming radiation (from the direction of the edge) is likely of the order of 1.667×10^{-4} that of light. With radiation transit times of 522 MY and 1.02 BY, respectively, one may calculate that NGC 7603 had to be 2.914 MLY from the edge at the time its light began the journey to Earth. For the more redshifted companion galaxy a figure of 2.835 MLY may be computed, which gives a separation of about 79,000 LY between the two systems. (It will be noted that a relatively modest change in our own position must be greatly magnified when translated into separation of two such interacting systems – thus imposing a strict limitation upon any permissible estimate of our proximity to the cosmic edge.)

Another rather interesting association consists of an open spiral galaxy known as AM 2006-295 (redshift 7,230 km/s) and its elongated companion (redshift 29,580 km/s), which appears to lie midway between (and behind) the nucleus and an arm of the larger spiral system. Once again assuming our Milky Way to be situated some 3 MLY from the cosmic edge, we may calculate the distances of these two galaxies upon the basis of light transit times of 483 MY and 1.82 BY, respectively. The figures which may be derived are 2.921 MLY from the edge for the main galaxy and 2.712 MLY for the companion. Hence, a separation of some 209,000 LY will be inferred to exist between the two systems. (It is to be noted that, without such an explanation, the measured angular displacement of the higher redshift member must imply a size which – in terms of the traditional velocity/distance scale – would make it unrealistically large!)

Upon the basis of a 3 MLY figure as our present distance from the cosmic edge, statistics concerning a number of other highly probable interacting systems (lying in the direction of the nearby edge) are as follows:

Where:
    R = implied redshift;
    T = time of light propagation to Milky Way; and
    D = distance from cosmic edge at time of light emission.

NGC 1232

Main Galaxy:
(R = 1,776 km/s; T = 107 MY; D = 2.982 MLY)

Companion System:
(R = 6,552 km/s; T = 393 MY; D = 2.935 MLY)

Inferred line of sight separation of galaxy pair is 47,000 LY.
(Traditional redshift separation is 287 MLY.)

Incidentally, a logical explanation is forthcoming for a second apparent companion to NGC 1232. Displaying a redshift of 28,000 km/s, this compact object would seem to shine through the disk of the nearby galaxy, which should serve to redden it considerably. To the surprise of many astronomers, it is observed to be quite blue in color! There is now every reason to suspect that it is indeed a distant background system which, due to pressure of encounters with vast swarms of D/R quanta, has had its light bent in such a manner as to project an image in front of the lower redshift galaxy's spiral arm. Most assuredly, this proffered explanation is able to resolve a mystery which must otherwise stand in flat contradiction to established precepts.

AM 0328-222

Main Galaxy:
(R = 1,800 km/s; T = 108 MY; D = 2.982 MLY)

Companion System:
(R = 19,700 km/s; T = 1.182 BY; D = 2.809 MLY)

Inferred line of sight separation of galaxy pair is 173,000 LY.
(Traditional redshift separation is 1.07 BLY.)

AM 059-4024

Main Galaxy:
(R = 6,730 km/s; T = 404 MY; D = 2.933 MLY)

Companion System:
(R = 16,400 km/s; T = 984 MY; D = 2.840 MLY)

Inferred line of sight separation of galaxy pair is 93,000 LY.
(Traditional redshift separation is 582 MLY.)

AM 2054-221

Main Galaxy:
  (R = 10,400 km/s; T = 624 MY; D = 2.898 MLY)

Companion System:
  (R = 46,860 km/s; T = 2.81 BY; D = 2.566 MLY)

Inferred line of sight separation of galaxy pair is 332,000 LY.
(Traditional redshift separation is 2.19 BLY.)

In actuality, the aforementioned figures must receive adjustment due to the influence of discrete D/R sources, which will serve to enhance all redshifts. For instance, the "repulsive" quanta that is constantly being emitted from our own Milky Way must impose a specific measure of retardation upon incoming radiation, increasing redshifts and changing statistics to a certain degree. Similarly, in the case of dominant galaxies with minor companions, the D/R factor will favor an increase in redshift of the less massive companion – leading to an overestimation of any deduced separation. (While necessitating some modification of figures, this is likely to be of relatively modest consideration when compared to the overall outward flowing of the universe, and we are thus enabled to obtain at least a "ballpark" idea as to the eccentricity of our cosmic location.)

Although a figure of 3 MLY may not be an unreasonable estimate of our present distance from the cosmic edge, it is desirable to examine other possibilities. (See *Appendix 1*, page 141.) A distance any greater than 5 MLY would seem to imply separations that are excessive for most systems, which might better be explained by separations of 1 MLY or less. On the other hand, in order to explain a separation as low as 69,600 LY, among components of AM 2006-295, it would take a very recent chance encounter not to show strong signs of gravitational disruption. (In point of fact, *both* of the twin arms of the higher redshift system do exhibit definite curvature toward the nucleus of the lower redshifted galaxy, which might indicate that this is indeed the case – implying that our Milky Way could actually be situated within a million light-years of the cosmic edge!) Additional study of such systems may be expected to reduce present uncertainty, which will be seen to hinge upon making an accurate assessment of how D/R quanta act to retard radiation with respect to galaxies of differing mass.

In view of this widespread "compression of images," toward the direction of the cosmic edge, it may well be asked why there are not more reported instances of discordant redshifts among cluster members. Quite simply, the answer must be that such instances do exist; but they are generally unnoticed because many redshift differences are actually so great that they appear to require membership in other galaxy clusters! It is only when close angular proximity is noted that the issue of anomalous redshifts is raised. (Also a contributing factor is the reluctance of many astronomers to seek evidence which would challenge the credibility of long accepted views involving the redshift/distance scale.)

Although a preponderance of discordant redshifts are to be found in the direction of the nearby edge, there are a number of anomalies – toward the cosmic center – which invite discussion. Perhaps the most impressive example discovered to date is that of the remarkable chain of galaxies known as VV 172 (R.A. 11^h 29^m, +71° 6' Dec.). While four members possess redshifts ranging from 15,673 to 16,252 km/s, the fifth system exhibits a rather discordant 37,062 km/s. Since visual observation would lead one to assume that they are all members of the same cluster, an explanation is clearly desired.

In this instance, we are probably dealing with only a modest degree of radial distortion, but a great deal of angular displacement toward the center of the universe. It is conceivable that the discordant member of this system has had its light bent in such a way as to cause it to appear as a fifth component in this unusual string of galaxies. In fact, it is to be expected that the phenomenon of a D/R force must produce many seemingly bizarre associations, including such publicized curiosities as Seyfert's Sextet and Stephan's Quintet – configurations defying a traditional interpretation, but which have a logical answer in terms of the "New Cosmology." Seyfert's Sextet, in which one galaxy member is revealed to possess a redshift almost 16,000 km/s higher than the others, may be viewed as a classic example of cosmic arcing so displacing an image as to create a spurious interloper.

Stephan's Quintet poses an intriguing example of such celestial deception. Located at somewhat less than a right angle from the cosmic edge, this group should really be considered edge inhabitants which have been subjected to considerable arcing. Three of these systems possess redshifts of 6,700 km/s; a fourth has been measured at 5,700 km/s; while the remaining galaxy exhibits a redshift of only 800 km/s. Added to the scenario is the presence of a large spiral galaxy known as NGC 7331, with a redshift similar to that of the low redshift member of the quintet. In turn, this major system would seem to have three small companions of redshifts

6,300, 6,400 and 6,900 km/s. When it is further considered that there exists a diffuse bridge of radio-emitting material between NGC 7331 and Stephan's Quintet, the entire picture begins to resemble a heavenly zoo.

Again, placing our Milky Way at a tentative distance of some 3 MLY from the cosmic edge, it is instructive to contemplate the following statistics for this peculiar configuration:

Stephan's Quintet:

NGC 7320
(R = 800 km/s; T = 48 MY; D = 2.992 MLY)

NGC 7318B
(R = 5,700 km/s; T = 342 MY; D = 2.943 MLY)

NGC 7318A
(R = 6,700 km/s; T = 402 MY; D = 2.933 MLY)

NGC 7317
(R = 6,700 km/s; T = 402 MY; D = 2.933 MLY)

NGC 7319
(R = 6,700 km/s; T = 402 MY; D = 2.933 MLY)

Inferred line of sight separation between:

NGC 7320 and **NGC 7318B**
49,000 LY (294 MLY)

NGC 7320 and all R = 6,700 km/s members
59,000 LY (354 MLY)

NGC 7318B and all R = 6,700 km/s members
10,000 LY (60 MLY)

Note: Separation, as implied by traditional redshift interpretation, is shown in parentheses.

The Enigma of Galactic Redshifts

Yet another instance of anomalous redshifts, which will be seen to have profound cosmological implications, involves galaxies of our own supercluster. Handicapped by a false model of creation, and in urgent need of a proper understanding of both gravitation and radiation propagation, astronomers have failed to recognize certain remarkable features which will become evident with little more than a cursory examination of galactic redshifts. Not only will a study of our Local Supercluster membership reveal numerous contradictions with regard to traditional views, but a proffered explanation is capable of lending strong support to several basic principles expounded in the "New Cosmology." There is also reason to believe that such a study will afford additional insight into our proximity to the nearby cosmic edge.

A most informative catalog of galaxy redshifts, depicting both distance and location, was compiled by astronomer R. Brent Tully. Published by Cambridge University Press (ISBN 0-521-35299-1) and entitled *Nearby Galaxies Catalog*, this comprehensive listing of some 2,367 galaxies forms the basis for a hitherto neglected study.

Proceeding upon premises inherent in the "New Cosmology," it is expedient to plot Local Supercluster galaxy redshifts as a function of distance and direction. Dividing the combined redshifts (in km/s) of a large sample group by their total distance (in LY), it is possible to make an interesting discovery regarding average implied velocity of recession per MLY of separation. With further subdivision into a number of bins relating to celestial location and distance, certain startling facts emerge which cast even more doubt as to the credibility of Big Bang philosophy.

Embracing a study of some 1,590 sample galaxies, essentials of this landmark analysis are summarized in Table 1. A mere glance is quite sufficient to disclose a truly astonishing pattern – one in which the average redshift (per MLY) is revealed to be considerably higher in the direction of the cosmic edge. (Indeed, such redshifts are seen to be fully 31% higher than those of galaxies occupying an identical volume of space in the opposite direction of the sky!) The overall picture is unmistakable and cannot possibility be negated by unfounded arguments of limited statistics. There is clearly a definite and symmetrical decline of galactic redshifts (per MLY) from the vicinity of the nearby cosmic edge toward the center of the universe.

Yet another truly remarkable observation is that of a pronounced reduction of center group redshifts from the "under 50 MLY" bin to the "50 to 100 MLY" bin. Involving quantities of 153 and 405 galaxies, respectively, the difference between these two adjoining bins is an amazing 55%! Needless to say, this sharp decline of redshifts is a fact demanding clarification.

Table 1. Average Redshift (per MLY) in Terms of Distance/Direction

Group	Under 50 MLY	50 to 100 MLY	Over 100 MLY
Edge Group Galaxies:			
R.A. 0,1,2,23h	25.25 km/s	25.10 km/s	24.69 km/s
-0° to -80° Dec.	[36]	[101]	[20]
Average Group Redshift [all 157 galaxies] = 25.02 km/s			
Adjacent Edge Group:			
R.A. 3,4,21,22h	23.48 km/s	24.36 km/s	24.17 km/s
-15° to -65° Dec.	[69]	[135]	[35]
Average Group Redshift [all 239 galaxies] = 24.16 km/s			
Center Group Galaxies:			
R.A. 11,12,13,14h	26.81 km/s	17.38 km/s	20.00 km/s
+0° to +80° Dec.	[153]	[405]	[140]
Average Group Redshift [all 698 galaxies] = 19.09 km/s			
Adjacent Center Group:			
R.A. 9,10,15,16h	20.02 km/s	17.94 km/s	20.61 km/s
+15° to +65° Dec.	[27]	[116]	[47]
Average Group Redshift [all 190 galaxies] = 19.89 km/s			
Right Angle Group:			
R.A. 6,7,18,19h	21.17 km/s	20.51 km/s	20.93 km/s
+0° to +80° Dec.	[107]	[103]	[104]
-0° to -80° Dec.			
R.A. 0,1,2,23h			
+25° to +65° Dec.			
R.A. 11,12,13,14h			
-25° to -65° Dec.			
Average Group Redshift [all 314 galaxies] = 20.81 km/s			

Note: Quantity within each bin is shown in brackets.

By coincidence, the cosmic center is seen to lie in the very same quadrant of the sky as the Virgo Cluster, a dense concentration of galaxies at the heart of our Local Supercluster. Thus it may be argued that gravitation acts to impose a measure of "infall" upon surrounding systems. While this might conceivably enhance the redshifts of some galaxies lying between us and the M87 Virgo group, and to reduce the redshifts of others located behind this massive cluster, gravitation alone cannot be the real story. This premise is borne out by the fact that most of the sample galaxies – between our Milky Way and the nearby cosmic edge – have quite similar redshifts (per MLY) to those residing between ourselves and the Virgo Cluster. Furthermore, as shown in Table 2, adjoining bins (extending 20 MLY on either side of the Virgo group) are affected differently! The bin closest to us is 9.74 km/s higher; while the opposing bin is only 2.87 km/s lower. If the force of gravitation is responsible, both bins should vary by the same amount. Clearly, a more subtle factor is involved and we must look elsewhere for a full explanation.

Table 2. Average "Center Group" Redshift (per MLY)

Distance, MLY	Quantity	Average Redshift, km/s
0 to 30	92	25.30
30+ to 50	61	27.91
50+ to 60 *(Virgo Cluster)*	226	18.17
60+ to 80	92	15.30
80+ to 100	87	17.93
Over 100	140	20.00

With the Virgo Cluster strongly concentrated within the 50 to 60 MLY bin, an informative picture emerges when we examine the adjoining bins which extend some 20 MLY on either side. If gravitation is the sole factor responsible for the noted redshift differential, then why is the closer of the two neighboring bins accelerated some 3.4 times that of the other? Clearly, gravitation is not the only ingredient, and a *nonvelocity* factor must be acknowledged. (It is contended that this missing influence is due to the interaction of radiation with D/R quanta. While a favorable flow can assist the propagation of such corpuscles, encounters with quanta ejected from our own Milky Way serve to retard approaching extragalactic radiation – thereby imparting a redshift which, in turn, may be defined as a measure of extended transit time.)

But problems raised by this redshift study go beyond peculiarities of implied motion and the need to consider revised views of gravitation and radiation propagation. Far deeper cosmological issues are surely involved. Since the Hubble flow rate must exceed any real recession velocity among the galaxies of our Local Supercluster, due to mutual attraction among the constituent members, a serious contradiction to Big Bang philosophy surfaces. The very fact that so many galactic redshifts are in excess of 25 km/s (per MLY) is strongly indicative of a Hubble flow of at least 40 km/s (per MLY). However, such a high rate would imply a Big Bang universe with an age/radius of no more than 7.5 BLY – a figure much less than the oldest stars inhabiting the globular clusters of our Milky Way! The alternative, of course, is to simply acknowledge either a different cosmological model or a nonvelocity component of many galaxy redshifts – *or both!*

In order to place matters into true perspective, it is advisable to ponder the origin of nearby galactic redshifts, as a mystery of sorts is seen to beg an answer. Specifically, why do so many nearby galaxies exhibit a redshift? How did they ever manage to acquire a redshift if the D/R factor is ineffective at a distance less than about 17 MLY, which is the point where repulsion is predicted to merely offset attraction? If held together by the force of gravity, one should expect that the random motions of nearby cluster members would serve to guarantee a roughly equal mixture of blueshifts and redshifts. Only when it is finally understood how an inordinate number of redshifts were produced may we properly address the intriguing problem of such galactic peculiarities.

The preponderance of redshifts over blueshifts is revealed to be positively staggering. Of the 2,367 galaxies listed in the *Nearby Galaxies Catalog,* only 19 are seen to possess a blueshift! In point of fact, 9 of these are due to rapid orbital motion about supermassive M87 and several large companion systems. (By way of confirmation, it is noted that these approaching galaxies are offset by an equal number of exceptionally high redshifted and receding systems.) The 10 remaining blueshifted galaxies are all situated within a few MLY of Earth and tend to be strongly concentrated more toward the nearby cosmic edge than in the opposite direction. In the face of this exceedingly biased redshift/blueshift distribution, it is little short of obvious that a *nonvelocity factor must be deeply involved in the vast majority of all Local Supercluster galaxy redshifts!* Indeed, such a conclusion is surely beyond question; it only remains to advance a credible theory to account for this observation.

In all probability, a solution is to be found in the interaction of radiation with the myriad swarms of D/R quanta so believed to permeate space. More specifically, it is the ability of our Milky Way to retard incoming radiation – through ejection of "repulsive" quanta – that is largely responsible for nearby galactic redshifts. (It has already been firmly established that, in every instance where two equidistant galaxies are of widely differing mass, it is always the smaller system which displays the higher redshift. In turn, this must lead one to conclude that a more massive galaxy is better able to facilitate the propagation of expelled radiation – presumably, through ejection of a stronger flow of "repulsive" quanta.) This being the case, it also follows that all galaxies possess an inherent ability to retard or slow incoming radiation – in the sense that the more massive the system, the greater the retardation.

Other than enhancement of redshift due to "cosmic arcing," and of interaction with clouds of interstellar hydrogen gas, it is logical to infer that the origin of nonvelocity redshifts is the outcome of two opposing influences. On the one hand, the propagation of radiation – between galaxies – is assisted by the flow of D/R quanta streaming outward from the emitting galactic mass. Conversely, all galaxies must act to slow approaching radiation through ejection of this same form of "repulsive" quanta. However, it is important to realize that the chief influence is bound to be the one imposed upon any incoming radiation. (This assumption is rooted in a combination of large scale scattering of D/R quanta over distance and small scale "focusing" of such quanta as it is condensed in proximity to a massive galaxy.) Hence, the noted "surplus" of galactic redshifts – toward the cosmic edge – may well have a ready explanation in the premise of extinction at the boundary of the universe, which acts to severely restrict the flux of inward flowing "repulsive" quanta. As a result, a superior outward flowing stream is enabled to impose a degree of retardation, or time delay. In the opposite direction, a stronger flow of "repulsive" quanta will permit our Milky Way to reflect this energy back toward the cosmic center (much as a mirror reflects light), which would account for the above average redshifts in the centrally located "under 50 MLY" group.

While it is tempting to ascribe the sharp reduction of center group redshifts (beyond 50 MLY) solely to infall toward the Virgo Cluster, a major factor is seen to be the strong flow of D/R quanta ejected from this same massive cluster, which acts to push radiation along with increased efficiency. Inasmuch as any retardation of radiation hinges upon both field strength and transit time, our Milky Way is less able to impart a non-velocity redshift upon more remote sources by reason of dissipation of D/R flux with distance. Hence, in conjunction with a reduction of flux due to expansion of the universe, systems located immediately beyond the Virgo Cluster may be expected to exhibit even lower redshifts (per MLY) until finally stabilizing at a rate more commensurate with the dictates of Hubble's law, exactly as observed. (In the direction of the nearby cosmic edge, where no such massive clumping of galaxies exists, a similar abrupt redshift reduction – beyond 50 MLY – is precluded by a scenario featuring a somewhat more homogeneous flow of D/R quanta.)

To the astonishment of the astronomical community, it has been reported that the redshifts of many members of our Local Supercluster are actually declining with the passage of time!* Through analysis of observational data, compiled during the course of the past decade or two, it is claimed that (in this interval) a reduction of typically 1 to 2 km/s is probably the rule rather than the exception. Needless to say, *upon the basis of traditional scientific beliefs, there is no conceivable explanation for what would appear to be a truly bizarre state of affairs!*

In light of theorems inherent in the "New Cosmology," it may well prove instructive to examine certain peculiarities which must arise in the vicinity of the nearby edge, where extinction serves to severely restrict the flux of "repulsive" quanta flowing back toward the cosmic center. In turn, this negation of D/R quanta will allow any inbound radiation to escape the normal repulsion exerted by our Milky Way, thus permitting inbound corpuscles to reach us in less time and with reduced redshift. (No perceptible change is to be expected in the spectra of very remote sources, as light transit times become overwhelmingly large in comparison to any reduction. Also, it is to be mentioned that the intense bending or arcing of radiation, which occurs at the extremities of the universe, assures that many "near edge" images will appear to be situated much closer to the cosmic center than to their true celestial coordinates.)

The question which now arises is whether nearby redshifts could actually turn into blueshifts. Clearly, the prospect of just such an occurrence, in the ridiculously brief span of only a few thousand years or so, is enough to raise eyebrows. This issue is likely to be at the cutting edge of future research and debate as astronomers, physicists and cosmologists desper-

---

* W.G. Tifft, *The Astrophysical Journal,* 382; pps. 396-415, December 1, 1991.

ately seek to resolve a paradox which is greatly compounded by erroneous views of gravitation and radiation propagation – to say nothing of a false creation model!

By way of coming to grips with the problem, it is advisable to ask one pertinent question. In short, how is it possible for nonvelocity redshifts, which take millions of years to arise, to be negated in but a small fraction of this time? Clearly, an explanation requires some sort of mechanism capable of producing a sudden and drastic impact upon radiation as it enters our Milky Way's sphere of influence. It must be evident, in fact, that traditional laws of physics are quite inadequate to resolve what (in the eyes of many scientists) must pose a puzzle to end all puzzles.

In an encouraging test of "New Cosmology" theory, a simple and most logical solution comes to mind. It has hitherto been presumed that both radiation and gravitational energy (including D/R quanta) are propagated at essentially the same velocity – namely, the speed of light. But what if "repulsive" quanta travel slightly faster than radiation? (The very premise that gravitation can flow from a black hole, while radiation cannot, is consistent with this hypothesis.) A difference of only a fraction of one percent can be shown to have a significant impact upon radiation that has been in transit for millions of years. Moreover, should D/R quanta be slightly less affected by cosmic arcing, this repulsive force will reach us by a more direct route and in less time. Together, this ability of "repulsive" quanta to outrun radiation (which is constantly having its speed adjusted through interaction with the infinitesimal quanta of space) will lead to the spectacle of a rapid decline of redshifts with respect to an Earth observer residing in the immediate vicinity of the cosmic edge.

To grasp the full picture of how this differential velocity could cause a sharp reduction of nearby redshifts, it is well to ponder matters in conjunction with the annihilation factor so believed to await us at the boundary of the universe. Since the last D/R quanta (emitted from a long-extinguished galaxy) will be seen to have passed us before arrival of its last radiation, it follows that the normal ability of the Milky Way to retard incoming radiation will dwindle rather abruptly – simply because our system will no longer have any "repulsive" quanta to eject toward such inbound corpuscles. As a direct consequence of this sudden cutoff of the means with which to slow approaching radiation, a situation must prevail whereby the nonvelocity redshifts of many local systems are reduced by reason of extinction at the cosmic periphery. Conceivably, a detailed study of the rate of change of neighboring redshift sources may one day afford a new and totally independent method of determining our proximity to this most portentous boundary – an opportunity which hinges upon accurate assessment of the inferred transit time differential between radiation and D/R quanta.

In conclusion, it may be said that this reported decline of nearby galaxy redshifts is exactly the proof needed to confirm essentials of the "New Cosmology," including the prediction that we presently reside in close proximity to an exceedingly fateful cosmic edge!

Nature has provided us with valuable clues concerning our status and location in a truly dynamic universe. It is to be hoped that the scientific community will not permit the sin of prejudice to cloud judgment, nor to restrict the quest for knowledge should this search threaten the credibility of established views. Most assuredly, any decline of local galaxy redshifts positively demands an urgent and unbiased investigation in the light of many new and viable principles.

The Microwave Background

In the opinion of many scientists, the observed microwave background radiation originated some billions of years ago in the intense heat and density of a Big Bang explosion. It is believed that expansion of the universe has served to reduce highly energetic radiation to the noted wavelength peak of about 1 mm which, in turn, has been translated to a temperature of close to 3°K. There is reason to suspect that this concept is both illogical and erroneous; in fact, it bespeaks an uncritical assessment of current nebulous views of radiation.

The very idea of stretching wavelengths – from the realm of high-energy gamma rays into the region of enormously weaker microwaves – poses a vexing contradiction with a basic tenet of physics demanding conservation of mass-energy. Ignoring the issue of particle mass, it fails to explain how such entities as electron/positron combinations (gamma rays) can incur a drastic reduction of mass simply because the universe is expanding. (Upon the basis of the "New Cosmology" every component of creation is seen to possess mass, or ability to commune gravitationally – a property which may be negated, as in the case of the elusive neutrino and radiation in general, solely due to extreme relative motion at the speed of light, since this is essentially the velocity by which gravitation is propagated.) It is thus inconceivable that cosmic expansion can have any effect upon the actual mass-energy congealed into corpuscles of radiation. All that it can do is to impose a reduction of flux and of frequency (wave crests per unit time) by reason of Doppler shift. However produced, the microwave background radiation cannot undergo any substantial change from time of formation to detection.

The low energy level and the remarkably isotropic nature of this phenomenon is suggestive of an entirely different origin than a Big Bang explosion. What is surely indicated is an homogeneous production mechanism, more characteristic of the *absolute zero* of outer space, than of the incredibly high temperatures of a primordial fireball. The most probable scenario to emerge involves elementary matter forms that are strewn throughout the expanse of space, where the chief source of energy which may be encountered is the larger "repulsive" quanta that is believed responsible for expansion of the universe. The problem is one of explaining how exceedingly minute quanta are absorbed and then emitted as more massive microwave radiation. In essence, the *microwave background reflects the average temperature of matter in the bitter cold isolation of extragalactic space!*

While a great deal remains to be explored in depth, it is logical to infer that (in the immediate vicinity of "maximum" particles) quanta of a certain size range will infrequently exist in mutual association long enough to enter into a degree of fusion. When the inevitable expulsion occurs, it could well be in the form of discrete corpuscles with masses characteristic of the microwave region of the spectrum. (It is to be expected that expulsion of absorbed quanta, in the almost absolute zero of space, will not be so abrupt as would be the case at a higher temperature.) Although only a small fraction of ingested quanta is likely to congeal into such radiation, what is produced will be propagated back and forth among components of the interstellar medium. Thus, a universal sea of microwave radiation (which just happens to be centered about the 1-mm wavelength) may be presumed to arise by reason of creation in the halos of isolated material particles, rather than originating in a superdense explosion.

While the vast majority of Big Bang proponents still cling to the belief of a fireball microwave origin, a growing number of reputable scientists have succeeded in presenting an impressive argument to the contrary. In spite of widespread refusal to recognize the idea of low wavelength radiation scattering by interaction with the extragalactic medium, there is strong evidence that this is indeed the case.* Not only are distant radio/microwave corpuscles effectively scattered in transit, to produce an ubiquitous celestial fog of low-level radiation, but this inferred dissipation of quasar radio emission readily explains an otherwise mysterious decline of such flux which is observed to occur with distance. (It has long been known that, in general, radio-loud quasars are relatively nearby and of more modest redshift.)

Since passage through a uniform field of microwave radiation is bound to produce a Doppler shift, extensive measurements have been made in order to determine the direction of our motion through space. It is reported that the temperature of the microwave background is slightly higher in the direction of the constellation of Leo; while it is less by an identical amount in exactly the opposite direction of the sky in the constellation of Aquarius. This has been interpreted to mean that the Milky Way, along with other members of our Local Group, shares a common motion (about 600 km/s) toward Leo. On the surface, this would seem to conflict with impressive evidence that our entire region of space is likely moving toward the nearest cosmic edge, which is believed to lie in the direction of Sculptor. What may be wondered is why observation appears to indicate movement in almost precisely the *opposite* direction!

To resolve this paradox it is essential to contemplate certain aspects of cosmic expansion in the light of the "New Cosmology." Since absorption and emission of quanta cannot occur in zero time, it is most logical to infer that a massive galaxy will resist acceleration more than minute corpuscles. Indeed, in order to impart motion to vast star systems, it is necessary for the impetus of "repulsive" quanta to be transferred from atom to atom – a somewhat less-than-instantaneous process as galaxy components are held together by gravitation. It must follow that this time lag will enable all outward-bound radiation to have a slight advantage over similar corpuscles propagating toward us from the direction of the nearby edge. In effect, any lag in our Milky Way's outward acceleration is tantamount to reducing the Doppler shift of incoming quanta (from the vicinity of the cosmic center) – thereby explaining why the microwave background radiation is observed to be a trifle warmer toward the center of the universe.

---

* See *The Big Bang Never Happened* (p. 277), by Eric J. Lerner (published by Random House, 1991; ISBN 0-8129-1853-3). See also the enlightening paper, *An Alternative Interpretation of the 3 K Radiation,* presented by a leading theoretical physicist, Professor Paul Marmet ("The American Association for the Advancement of Science," San Francisco, June 1994).

New Horizons in Cosmology

The small discrepancy between microwave observations and quasar distribution, as a means of determining our orientation with respect to the cosmic center and nearby edge, would appear to have a ready answer: it may well be ascribed to residual orbital momentum stemming from the Small Bang which gave us birth! It would, in fact, be surprising if no vestige at all remained of the strong initial rotary motion of matter forms about a supermassive black hole creation site. On the whole, one must concede that there is a most remarkable agreement between these two methods of deducing our motion toward the nearby cosmic edge.

In negating this last remnant of support for a Big Bang cosmology, it is clear that evidence against this model is now so overwhelming that a parallel exists with the state of affairs which prevailed during the pioneering days of Copernicus, Kepler and Galileo. It is to be hoped that science will soon recognize the futility of attempting to preserve an outworn concept and begin to explore new horizons. The rewards will surely exceed all expectation.

5

Cosmic Reincarnation

The Great Cosmic Pyramid – The Universal "Time Constant" – Beyond Human Evolution – A Cosmic Paradox – Summary

Both theoretical and observational evidence would appear to lend support to the concept of a bound and finite physical universe. But many questions remain unanswered. Not the least of our problems is to devise a formula linking cosmic expansion with the *Motivation* behind creation. Furthermore, it must be conceded that the very existence of an enclosed system does not preclude the real possibility of *other* universes! However, restricted as we are by limited insight at our present stage of evolution, it is perhaps prudent to reserve such comment to that which is conducive to investigation. In so doing, it will be shown that the physical cosmos is characterized by a rather unique series of mathematical relationships – a feature which would seem to indicate that, for all practical purposes, *our own universe must constitute a bound system sufficient unto itself!*

The Great Cosmic Pyramid

A most instructive picture of the universe emerges as a result of scientific deduction and logic. At the lower extremities of the Cosmic Pyramid, the intrinsic worth or value of a particle of matter is revealed to be directly related to the quantity of "congealed energy" fused together into a common body – an entity that is as one with regard to such properties as space and time. Beyond the realm of atomic nuclei, with their enormously enhanced complexity and status in comparison to an elementary "**A**," we soon encounter a revolutionary stage of expression in the form of organic life.

This new level is characterized by a marked transformation, in the sense that fusion of larger quantities of "**A**" no longer serves to produce a corresponding increase in the mass of an entity. At this point any further "desire" must be manifested in some other manner and upon an entirely higher plane. For instead of expending energy as the phenomenon of gravitation, in an endeavor to attract mere matter which now lies below its status, such effort is directed to the process of achieving maximum interaction and harmony with similar members of its own world. Subsequently, as organic evolution progresses upward to the status of man at his pinnacle, so it may be inferred that these higher levels of spirit must represent fantastic amounts of energy since contained and converted to increased segments of God's Nature. A superior cosmic status, being the product of *deserved* fusion, may only be obtained with the passage of time and at the expense of a great deal of effort or energy, as we are often quite aware.

Even the strange paradox of Eternity begins to shed some of its aura of mystery once the property of time receives interpretation as a relative experience – one that will surely vanish among members should interaction reach the point where it becomes instantaneous. To the extent that a degree of reciprocal communion is achieved, so there will no longer exist any reason for separate identities, and there will be a loss of previous personalities as they become united into fewer entities of greater worth. Time may thus be viewed as a unique attribute of *imperfection,* prevailing at all levels of creation below that of God, but having no substance or meaning once Perfection is reached – a state which might well be described as *Total Communion* or *Instantaneous Motion.* Eternity or Timelessness, rather than the very transitory world of space and time, is disclosed to be the Ultimate cosmic state. Just as surely as time appears at the lowest strata of creation, so it will cease to have any meaning at the apex of the Great Cosmic Pyramid once spirit is deserving of being united into One Harmonious Whole.

Viewed upon the large scale, the rate by which spiritual fusion occurs throughout the universe cannot possibly be ascribed to a random factor. It is clearly a phenomenon in which advancement at one level is intimately connected with progress achieved at all other levels. The equilibrium of the cosmos can only be maintained if spirit of the lowest caliber is created at a rate capable of balancing evolutionary movement to higher levels.

Similarly, the frequency of fusion must facilitate a steady upward flow of spirit, free from the catastrophe of disruption that would arise should either a shortage or surplus be produced at any one level.

Notwithstanding this necessity to preserve overall stability, it is important to realize that – on a short-term basis – there need not be an ironclad rule insisting that actual instances of physical fusion must always take place, over the entire range of cosmic steps, with unswerving regularity. On the contrary, it is sufficient only that adjustments intercede periodically to restore long-term balance and justification. Judging by the relative abundance of the many physical body-forms, housing a wide range of spiritual complexes, it is evident that phases of both literal fusion and quasi-fusion exist side-by-side in the universe.

Although a pattern of literal fusion (with reduction of numbers) may be inferred during the evolution of "**A**" to the level of atomic particles, a certain complexity surrounds the involvement of animate structures. A vast gap seems to prevail between the numbers of the material particles and elementary life forms. This discrepancy must surely indicate that, upon occasion, a great many spiritual entities may undergo delayed fusion into one radically different and higher structure. Prior to this significant event, justice could be rendered in a quasi-fusion state. In a sense, spirit may enjoy the "fruits of fusion" without a literal fusion having transpired. At the "pre-life" stage this condition is probably fulfilled within the extremely hot and dense interiors of stars, especially degenerate stars where acts of reciprocal communion must be so magnified as to simulate actual fusion.

The course of biological life appears to be characterized by a mixture of fusion and quasi-fusion. In addition to a reduction of body-forms with status there are also instances where the advantages of fusion – without fusion – are readily available to many life forms. Man, for example, affords an excellent illustration of the benefits accruing from the ability to commune with other members of one's own species. By reason of this opportunity to cooperate and share in a wealth of knowledge and experiences, his environmental status is seen to be enhanced to an extent not otherwise possible.

This leads us to ponder the long-term prospectus of mankind as a physical entity. Quite regardless of any inferred future reduction of population with evolutionary progress, it is inevitable that man's present quasi-fusion era must continue to alleviate instances of any spiritual injustice until his planet's final demise. What then? Indeed, since it is rather inconceivable that man as such could ever expect to conclude his evolution toward Perfection without there most assuredly being some transitionary stage, we may well ask the nature of this new and highly revolutionary structural form.

Strangely enough, his destiny could very conceivably be found in the stars! If this should seem slightly melodramatic, we have merely to consider the unique properties of a fully degenerate star, where matter is so compressed as to virtually constitute a *super atomic particle!* Not only can we point to the incredible density and mass of a neutron star (about half a million Earths compacted into a sphere only 10 miles or so in diameter), but the stellar black hole state of matter may be described as a total collapse and absolute fusion of a slightly greater segment of creation.

Becoming manifest as a spiritual entity of tremendous worth, once a critical stage of contraction is achieved, yet another quasi-fusion era may be initiated as a star of superdense proportions continues its development upon a new plane of existence. One can but speculate as to the details involved in subsequent evolution. Presumably, progress is facilitated by means of establishing reciprocal communion with similar entities – *interaction of a nonphysical nature and no longer restricted to the velocity of light.* Finally, upon attaining the ability to express Total or Instantaneous Communion, time will cease to exist for such advanced entities and a *literal fusion into God may be inferred!* Somehow, as a consequence of *black holes* concluding this momentous transition from the physical world of imperfection and time, the surplus "energy" so released is manifested as a new generation of "**A**" within the confines of a Cosmic Creation Zone.

By way of illustrating the close interrelationship between cosmic extremes we might point to five rather curious numerical associations embracing phenomena of the ultra-small and the ultra-large. Involving the immense sum of 10^{40} (1 followed by 40 zeros), these similarities may be stated as follows:

1) The electrical charge of a positive/negative pair of material particles, at really close range, exceeds their mutual gravitational influence by a factor of very nearly 10^{40}.

2) The density ratio between highly degenerate stars and the average density of matter throughout the expanse of the universe is of the order of 10^{40}.

3) Stretched out in a straight line, like a string of beads, it would take about 10^{40} proton nuclei to reach the distance at which the galaxies are receding at the speed of light.

4) The repulsive aspect of gravitation, at a distance equal to the diameter of a "maximum" particle nucleus, is seen to be some 10^{20} times weaker than attraction — in essence, strength of repulsion is the square root of attraction.

5) The square root of the number of protons, within the expanse of our "observable universe" (10^{80}), is none other than 10^{40}.

In each instance these relationships may apparently be looked upon as an expression of fusion versus non-fusion. Not only do they tend to substantiate the underlying cosmic principle of fusion, but there is even reason to suspect that a valuable clue lies therein as to the number of "**A**"s fused within the level of a "maximum" particle.

It will be recalled that gravitational influence, in the final analysis, had to be defined as an inborn desire for fusion on the part of all material components of creation — a universal expression of intrinsic worth dissipated over the depths of space. Essentially, it is a reflection of a particle's ability to achieve attraction by means of communion with the physical universe as a whole. In contrast, the electrical force of a particle is a manifestation of this same worth condensed to a volume of space equal to that of its immediate self: the halo of energy surrounding the material body in question. Since there is good reason to believe that the value of "**A**" fused within a proton or an electron does not exceed that of the halo by a really exorbitant figure, it follows that exposure to another such charged particle will produce a magnified reaction that is *proportional to the the quantity of "**A**" involved!* Hence, it is logical to infer that a "maximum" particle likely consists of roughly 10^{40} "**A**"s. In a sense, the electrical force may well be considered an indication of achieved fusion; whereas gravitation may be viewed as a sure sign of potential fusion associated with Destiny.

With regard to the second similarity, a fully degenerate star must represent a state of communion tantamount to "total" density or fusion into one overall entity. On the other hand, the average density of matter in space may be defined as the "minimum" degree of interaction that is permitted to exist without requiring new creation. Within the Cosmic Creation Zone, the introduction of "**A**" will be inferred in all regions falling below average density, in contrast to those forms that are able to achieve critical condensation leading to a degree of total fusion. In essence, these extremes may be thought of as depicting maximum and minimum concentrations of energy within the framework of a bound system – one in which the velocity of light denotes a total separation among components of creation.

The connection between the diameter of a proton nucleus and the size of our universe is, of course, directly linked to the issue of average density and hydrogen abundance. Starting with the fused state of such a particle, it will be seen that this overall distribution is consistent with a twofold increase in distance (an eightfold increase in volume) for each fourfold increase in the number of proton nuclei. Restated slightly, the pattern is one of a fourfold decrease in cosmic density with each doubling in the quantity of matter residing within the universe. To extend the picture, it is noted that a fourfold increase in mass is capable of expressing the same strength of any gravitational communion over eight times the volume of space. On the surface this is possibly surprising, since it might be wondered why the two should not coincide. What mysterious factor enables the force of gravitation to permeate as much space as it does?

The explanation presumably hinges upon a proper assessment of the roles played by time and evolutionary processes. For every increase in the creation of matter, so must it add to the task of achieving full communion in a universe where interaction among material bodies is regulated by the velocity of light. Should we proceed upon the premise that every level of spirit is able to double its status in unit time, it may be deduced that each successive level must possess the power to attract twice the volume of "**A**." But since a larger amount of matter must lengthen the time required to unite creation, such an addition can only take the form of increased separation or distance – essentially, a disguised aspect of time or imperfection. The ensuing pattern of isolation, which would preserve equilibrium by fulfilling the basic requirement of each "**A**" being able to obtain fusion with another "**A**" in the same interval of time, is one of a fourfold increase in matter with every doubling of distance. Upon this basis a twofold decrease of density – beyond what would be expected to characterize a simple doubling in the quantity of matter – will be seen to accompany each "step" in the Cosmic Pyramid. For not only must spirit be unsatisfied to maintain the status quo, but it is motivated to seek its Ultimate Destiny through further fusion. Hence, in pondering the ability of gravitation to spread its influence over space, we are very likely witnessing a graphic demonstration of spirit striving unceasingly to effect a *future* doubling of status.

A philosophical basis for the inferred D/R factor would thus seem implicit in theory. It is surely more than a coincidence that, at the point where this subatomic repulsive force first becomes manifest, it does so as the square root of the number of fused "**A**"s which are now believed to constitute the level of a "maximum" elementary particle. Indeed, it is essential that a twofold increase in the strength of any repulsive force induce a fourfold increase of distance in order to achieve a balance between the rate of new creation and the rate of cosmic evolution. In effect, it is the rate of spiritual advancement which determines the time scale of Hubble's constant!

With a total of 10^{80} protons comprising our physical universe, and 10^{40} fused "**A**"s per proton, it will be deduced that Perfection – or any one "level" within the Great Cosmic Pyramid – is equivalent to some 10^{120} "**A**"s. In terms of spiritual evolution, this Ultimate State would involve approximately 400 "steps" or occasions of doubling. As a result, an entity of the status of a proton is computed to reside 1/3 of the way toward the top of the Pyramid, having already climbed some 133 "steps" from its origin as an infinitesimal "**A**." Upon this premise, slightly less than 1% of all the mass-energy in the universe is likely to be incorporated into actual "maximum" particles.

Such an assumption can be put to an interesting test. What may prove to be so highly significant about this figure is the excellent agreement with certain cosmological considerations as they pertain to the Steady-state model of creation. According to this principle there are very definite limitations as to the relationship between average density of matter in space and the rate of cosmic expansion. With Hubble's constant established at roughly 12 billion years, no more than a few percent of all the material in the universe may be so concentrated as to form galaxies without threatening stability. Were a substantial proportion locked up inside stars it would contradict the assumption of Continuous Creation, which demands a preponderance of newly created matter forms in space. Conversely, too great an abundance between star systems – with Hubble's constant being what it is – could lead to the spectacle of our gravitationally weak Local Group of galaxies possessing a lower-than-average density – thereby causing the cluster to expand apart, which it is not doing. On the whole, our inferred ratio of mass-energy to actual matter, within the expanse of the universe, is seen to afford a most remarkable agreement with observation.

The Universal "Time Constant"

Certainly a most intriguing prospect now awaiting us is that of determining the universal *"time constant,"* which must surely regulate the entire process of spiritual fusion. For it is an integral feature of theory that all levels of spirit, within the Great Cosmic Pyramid, must succeed in doubling their worth in some common interval of time. A solution to this problem will make it possible to derive a useful picture of the average time that an individual must serve in order to achieve the equivalent of one doubling of status. Moreover, it will also provide insight into the highly relevant question as to when one may expect a future rebirth.

Evaluation of this time aspect entails consideration of certain factors. For instance, the rate of disintegration (at the edge of the universe) must be included in any formula attempting to express events in terms of a human lifetime. Due to a steady annihilation of worlds containing a variety of life forms, there will prevail a situation in which all disrupted levels of spirit are compelled to seek rebirth on appropriate planets still residing within the Cosmic Creation Zone. Along with other factors pointing toward a similar conclusion, this introduces the suspicion that, in terms of the planet with which a spirit was last associated, considerable time could elapse between reincarnations. In other words, biological evolution is likely to reflect a "sharing" of available body-forms with displaced spirit from alien worlds – with the result that spiritual entities are, in effect, excused from having a physical manifestation over most of the course of a planet's lengthy history! (Time, of course, simply does not exist for spirit between incarnations, just as it appears to cease when we fall into a deep and dreamless sleep.)

Thus, in theory, it could follow that those who manage to increase their intrinsic worth in excess of the average will receive what might well be equivalent to a longer postponement in their next rebirth – thereby enhancing prospects of securing an existence in more suitable surroundings. On the other hand, one who has made relatively little or no progress (whether or not restricted by premature death) may expect to be reborn in but a fraction of this time and, if it should be so justified, in a less favorable environment were a regression involved. Perhaps, in this manner, nature permits us some latitude in determining the circumstances into which we are reborn – including the specific moment of rebirth within an evolutionary era. Invariably, there is a great advantage to be gained from being reincarnated in a more advanced and civilized age.

A truly great challenge, confronting the inquisitive mind of mortal man, is to deduce the *"time constant"* of spiritual fusion. For without a steady and orderly upward flow of spirit, toward an Ultimate State of Perfection at the apex of the Cosmic Pyramid, the universe would be totally lacking in both Purpose and Justice. In spite of the many uncertainties that are encountered when pioneering into virgin territory, a solution of the time/evolution relationship is a task which cannot be avoided if true knowledge is to be our goal. While admittedly posing a structure of high complexity, our universe is not without numerous clues and much vital information – information that would have been treasured by many wise scientists and philosophers of the past. If complete success is not to be within immediate grasp, it should never be because we lack the fortitude to try.

Clearly, the *rate of expansion* of the universe must constitute a key factor in any approach to the time aspect of cosmic evolution. Why does the universe double its size in an interval of close to some 12 billion years? Why this particular figure? Obviously, the greater the rate of expansion, the greater the degree of disintegration at the edge of the universe. Inasmuch as this annihilation factor must be balanced through creation, it will be seen that the introduction and evolution of new "**A**" is closely related to the recession velocity of the galaxies. The issue, therefore, is basically one of determining precisely how "**A**" is introduced with respect to each cycle of Hubble's constant.

Upon due consideration, the one logical possibility would seem to embrace the premise that *one entire level of new "**A**" is created in just such a span of time and is able to fulfill its Ultimate Destiny by evolving to a state of Perfection in this same interval!* Should this be the case, it must follow that the time required by spirit to double its worth may be determined by dividing Hubble's constant by the number of "steps" or doublings separating "**A**" from Perfection.

Assuming a proton to consist of 10^{40} "**A**s," and a physical universe of 10^{80} protons, we have a spiritual pyramid of some 10^{120} "**A**s" per level and a total of approximately 400 "steps." (In view of the most unique relationship of the sum of 10^{40}, as it relates to the size of our finite universe, along with the inability of material forms to communicate at velocities exceeding that of light, it is perhaps in order to treat such a region as though it constituted a complete and closed system terminating in what – for all practical purposes – we may define as God.) Subsequently, a period of roughly 30 million years per occasion of doubling may be derived.

Proceeding upon this basis, it will be deduced that only 1/400 of an entire level of new "**A**" will be created in a time span of some 30 million years. In effect, this means that only one spiritual entity in 400 may have a "conscious" (or physical) manifestation at any given instant. But if 30 million years is the allotted time, in which one entire level of spirit must double its status, this very restriction will introduce an additional factor of 400 into the time permitted for actual instances of physical incarnation. Dividing our 30 million year period by this further factor of 400, we are left with about 75,000 years as the average elapsed time for such a "conscious" existence between doublings.

Yet another factor is imposed by disintegration at the edge of the universe, as displaced spirit must periodically be reborn within the confines of the Cosmic Creation Zone. Since it takes time for new stars to form and for planets to evolve biological entities, it is evident that this factor is very much related to status. Calculation of the average age of matter, within the framework of the Steady-state model, yields the information that only about one part in 20 will be as old as Hubble's constant.

Upon this line of reasoning the average time required by spirit – as a "conscious" entity – to double in value may be expressed by the formula:

$$T = \frac{SH}{DLS^2} \quad or \quad T = \frac{H}{DLS}$$

Where:

- T = time required to achieve one doubling of status in terms of a "conscious" existence;
- S = total number of cosmic "steps" from "**A**" to Perfection (or Full Communion) within a finite physical universe some 18 BLY in radius;
- H = Hubble's constant (deduced to be about 12 BY);
- D = disintegration factor at the edge of the universe in one cycle of Hubble's constant; and
- L = level of spirit in question – defined in terms of the number of "steps" evolved beyond the initial created state of an infinitesimal "**A**."

Assuming that 400 "steps" extend between "**A**" and Total Communion, it now becomes essential to determine the position of mankind in this overall scheme. With a hydrogen atom established at "step" 133, this problem may be resolved by a comparison of the relative abundance of the proton to man throughout the Cosmic Creation Zone.

By way of assessing this ratio, it may be estimated that about 10^{22} Earth-type planets presently exist in the universe, of which roughly about one in 10^5 will currently feature biological evolution equal to man at his present level. Multiplying our estimated 10^{17} planets by the sum of 10^{10} (representing an average long-term population of man per planet), we arrive at a total of 10^{27} human beings at this one particular level. Since some 10^{80} protons constitute our physical universe, we may deduce a proton-to-man ratio of about 10^{53}:1. Translated into "steps" in the Cosmic Pyramid this sum is equivalent to some 177 doublings. Accordingly, the level of man may be derived by adding these 177 "steps" to the 133 "steps" believed characteristic of a proton – leading to the conclusion that man presently resides at about "step" 310.

With this information, we may now attempt to resolve the "time constant" for man as per our formula. Therefore:

$$T = \frac{12,000,000,000}{20 \times 310 \times 400} = 4,800 \text{ years } (approximately)$$

Where:

T = time required to achieve one doubling of status in terms of a "conscious" existence.

Thus, it would be calculated that, upon the average, man must serve a total of something like 4,800 years (or almost 70 lifetimes of 70 years) in order to double his intrinsic worth. (Any uncertainty in computing the ratio of man-to-proton will not seriously alter our estimate, since an error by a factor of a million will only make a difference of about 6%, and a million million error only a 12% difference!) With such insight into man's status in the Cosmic Pyramid, along with knowledge of the "time constant" of spiritual fusion, we may now face the issue of reincarnation in terms of Earth evolution. During what past age did we last live? Moreover, in what future era are we likely to be reborn?

Considering the necessity of accommodating displaced spirit from other worlds, along with the overall cycle of doubling which is of the order of 30 million years, reincarnation may essentially take the form of a projection well into the future and into quite another planetary environment. Thus, if man requires 4,800 years as a "conscious" life form to achieve one doubling of status over a period of 30 million years, it will be tantamount to an elapsed time of almost 6,200 years for each year of a past life. Based upon an average life expectancy of 70 years this would call for a rebirth some four to five hundred thousand years into the future. However, if only 1/400 of an entire level of "**A**" is created per 30 million year cycle, in terms of Earth evolution the *apparent* elapsed time will be less than the *actual* time by this very same factor. Subsequently, for all intents and purposes, the reincarnation of man may be linked to the basic formula:

$$R = \frac{YH}{TS^2}$$

Where:
- R = rebirth of man (in years) in terms of a future era on Earth;
- Y = the number of years lived during last incarnation;
- S = total number of cosmic "steps" from "**A**" to Perfection; and
- H = Hubble's constant.

Hence:

$$T = \frac{70 \times 12{,}000{,}000{,}000}{4{,}800 \times 400 \times 400} = 1{,}100 \text{ years } (approximately)$$

Where:
- T = time required to achieve one doubling of status in terms of a "conscious" existence.

In other words, what probably transpires is that man, in relation to the environmental status with which he was last associated, is reborn into an evolutionary era only a thousand years or so into the future.

Beyond Human Evolution

Upon contemplating the future of mankind, as a biological species, it may be theorized that each additional doubling will require some 75,000 years in terms of Earth evolution. The eventual passage of man, beyond his present transitory and rather frail human form, could very well have a solution in the astronomical curiosity of a *celestial black hole!* (Neutron stars, which may in time evolve into stellar black holes as gradually declining rotation permits still further contraction, could also share such an exalted status.) Best described as immense concentrations of mass-energy contracted into a state of virtual-spatial oblivion, the ratio of mankind to this unique state of fused matter should allow computation of the number of "steps" needed to achieve the inferred evolution. If this is indeed the case, it may in truth be said that man's destiny is in the stars!

Strange as it might first appear, there are some very convincing arguments for linking man's future to a phenomenon that would tend to tax the imagination. Not only does this state of "maximum" density meet all the required attributes of complete fusion – thereby creating a spiritual entity of a specific level – but an excellent agreement of numbers can be derived which would connect theory with observation. It is surely more than a remarkable coincidence that an estimate of all the stellar black holes likely to exist in the universe should turn out to be equal to the probable number of Earth-type planets that are allowed to fulfill their evolutionary roles. Moreover, it can be shown that the quantity of "**A**" fused within a typical stellar black hole (some 10^{58} protons x 10^{40} "**A**s," or about 10^{98} fused "**A**s"), when multiplied by the number of such objects in the entire universe, is very likely of the same order as the total spiritual worth (or "**A**" content) of one complete population of Earth-man multiplied by the quantity of all other worlds of a similar evolutionary level. The implication is little short of obvious: *A stellar black hole may well represent the eventual destiny of a humanity, so deserving of intimate harmony, as to become fused into a new entity of truly enormous magnitude!*

By way of determining the number of cosmic "steps," which exist between present-day man and the lofty status of a stellar black hole, it is necessary to examine what transpires as man finally evolves beyond his current level which we might define as **M**^{1}. Upon doubling his worth, and attaining the status of **M**^{2}, it will be noted that a gap has suddenly appeared in our planet's otherwise unbroken chain of evolutionary life forms. In effect, there is no longer an **M**^{1} level. Similarly, after **M**^{2} progresses to the status of **M**^{3}, the gap will widen to include the absence of both the **M**^{1} and **M**^{2} levels. For the first time in our planet's history *a gap has occurred in the very evolutionary continuum* – one that will grow steadily in size until man's eventual extinction as a biological entity. Assuming that only one Earth-type planet in 10^5 will feature human life equivalent to a specific "**M**" level, it may be computed that about 17 doublings will suffice to accomplish the inferred evolution from **M**^{1} to the status of a stellar black hole.

Accordingly, this momentous stage may be looked upon as constituting "step" number 327 in a Cosmic Pyramid of 400 "steps."

The hidden significance of this 17 "step" gap, in the evolutionary sequence of all Earth-type planets, lies in its ability to resolve a potential paradox with regard to a final merger of "**M**"-type spirit into a stellar black hole. Without the existence of this gap, which must result in stellar black holes outnumbering any one specific "**M**" level by a factor of 10^5, we would encounter a somewhat outlandish situation at the more advanced stages of man's evolution. Since theory demands a reduction of numbers in direct proportion to higher status, it follows that the final "**M**" stages would otherwise be quite literally populated by a mere handful of individuals. The inferred gap of some 17 "steps" will greatly alleviate this problem by assuring that the ultimate "**M**" stage, immediately prior to achieving stellar black hole fusion, will consist of not less than 10^5 spiritual entities. When combined with the ratio of 10^5 stellar black holes in excess of a solitary "**M**" stage, we arrive at a total of 10^{10} entities – exactly the sum needed to account for a typical $\mathbf{M}^1$ population level! Hence, it is no wonder that there is a curious similarity in the numbers of stellar black holes and Earth-type planets. It does, in fact, seem to necessitate an apparent "surplus" of 10^5 stellar black holes in order to balance cosmic books in terms of "**A**" content.

To grasp the full picture of this crucial relationship, involving relative abundance and fused "**A**" content of the various cosmic levels, it is perhaps constructive to assess the total range of "**M**" stages in some detail. (See *Appendix 2*, page 142.) Immediately prior to our present $\mathbf{M}^1$ stage, it will be noted, the abrupt increase (10^5) in worlds must require a corresponding reduction in the number of pre-human entities. This is precisely what is observed when we examine the distribution of our own terrestrial life forms. The comparative shortage of primitive man is clearly no accident; nor is the recent sharp increase in human population an unprecedented occurrence. In both instances we are likely witnessing nature's balancing of appropriate cosmic equations. (Incidentally, the relative scarcity of early man tends to support the conclusion that there is really little difference in spiritual value between the higher animals and primitive man – with the family pet possibly only short steps away on the evolutionary ladder!)

In predicting the future of mankind, as he passes through the successive "**M**" levels, it is to be inferred that each doubling will involve about 75,000 years in terms of Earth history. After all 17 "**M**" levels (or generations of human doubling) have been attained, it may be calculated that some 1,275,000 years will have elapsed as man finally concludes his evolutionary saga on the planet Earth. From this point onward our world will continue to feature all present life forms below that of M^1; but any trace of a higher intelligence will likely be conspicuous by its absence. In actuality, *man will have vanished from the face of this planet simply because he succeeded in evolving beyond the human form!* (This deduction would readily explain any difficulty in contacting intelligent life forms on other planets. It is not so much that they are intrinsically scarce in our Galaxy, as it is that large numbers of such civilizations have already come and gone in the relatively short time span allocated for human supremacy.)

Invariably, it is the evolution of the human brain itself which provides the mechanism for his eventual reduction in numbers. Along with the degree of spiritual progress achieved during a past life, the rebirth of a soul must be regulated by two features. On the one hand, we have the quantity of spirit awaiting incarnation; while on the other, we have the number of suitable body-forms available for housing a specific level of spirit. With evolution serving to very gradually change the structure and caliber of man's brain, the incorporation of human souls will become highly restrictive as fewer entities of a particular status seek manifestation in fewer appropriate body-forms. Essentially, nature will be seen to impose its own foolproof method of birth control. As man pursues his ascent of the Cosmic Pyramid, so emphasis will be shifted from quantity to quality. (See Figure 12.)

Beyond the exalted status of a stellar black hole lies another universe embracing new and unfamiliar laws. Within this lofty domain of *Principle,* spirit is enabled to climax its lengthy evolution by transcending such mundane attributes as gravitation, electricity and time. In the eyes of mortal man the black hole state may very well be worshiped as a wondrous God-like realm of reciprocal communion – from whence Perfection will ultimately ensue as Total Fusion of Black Hole Minds becomes a reality and time ceases to exist! Somehow, as a direct consequence of this momentous transition from imperfection and time, the mass-energy so released is manifested as a new generation of "**A**" within the confines of our physical universe.

Perhaps, of all matter forms subject to instant disintegration at the edge of the universe, the unique properties of black holes could suffice to preserve them from sharing the same abrupt fate. Such an exception might even turn out to be both a logical and a necessary inference – especially in the case of the more massive quasar-type black holes. Ranging in size from several solar masses, for a typical stellar black hole, it is conceivable that a chain of successively larger singularities could exist (at the centers of globular clusters and galaxies) up to some billions of times the mass of our Sun. If the supermassive black hole at the center of the giant galaxy known as M87 (about 5 billion solar masses) is any indication, this would seem to represent the upper limit for the size of a black hole. Simple calculation of

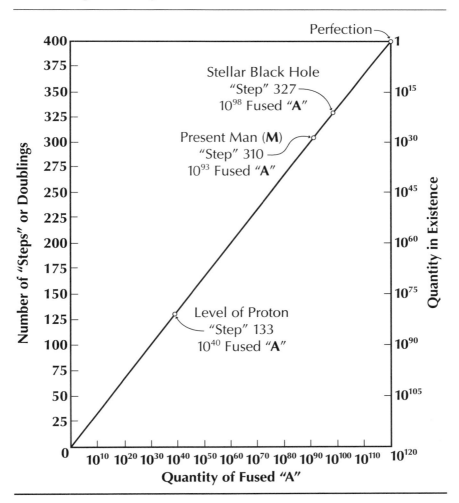

Figure 12. **Pyramid Structure of the Universe**

the "**A**" content, for an object of this nature, leads to a total of something like 10^{108} fused "**A**s." For an entity of this evolutionary status, its theoretical abundance appears to be much higher than what is observed. This discrepancy is promptly removed, however, by the simple expedient of exempting black holes – at least supermassive quasar-type entities – from any premature extinction at the cosmic periphery.

In any event, a truly amazing picture emerges as to what lies beyond the human form. We are at last in a position to glimpse a future that must exceed our wildest dreams. Upon contemplating the premise of an Eternal Universe, it may well be said that mankind now stands at the threshold of an entirely new and promising field of inquiry, the surface of which has barely been scratched.

A Cosmic Paradox

While a physical universe of finite size is thus theorized, and with stability assured by reason of an annihilation factor equal to that of creation, a paradox arises when we contemplate a connection with the spiritual aspect of cosmic evolution. If spirit forever rises from an initial state of a lowly "**A**" to one of Perfection (or God), then may it not be said that – with the passage of time – God is becoming increasingly more Omnipotent? Conversely, might it also be argued that God was less of an Entity in the distant past? The issue of a finite physical universe is therefore seen to differ from the issue of spiritual evolution in one crucial respect. In essence, our physical environment may be likened unto that of a catalyst – a medium remaining unchanged in the long run while facilitating an endless progression of spirit up the Great Cosmic Pyramid.

Conceivably, a solution to this Eternity paradox may reside in a property so believed to characterize the Ultimate State of Perfection – namely a state in which *time* has literally ceased to exist! In the absence of time – where communion is instantaneous – distance and physical dimensions become meaningless terms. Devoid of both time and size, the problem of an ever-increasing and evolving Creator tends to fade into oblivion. Likewise, the property of quantity – being linked to size – is similarly shrouded in nebulosity should there be no such thing as time. Looked at from the point of view of mortal man there does indeed appear to be a contradiction. At the level of Perfection, human logic and reasoning may very well have no substance.

It is unfortunate that our rather limited faculties prevent us from presenting a more satisfactory answer to this "Cosmic Paradox." However, it is also clear that one who is less than Perfect must be precluded from knowing all. At our current evolutionary level it is, perhaps, sufficient to recognize that any distortion of Reality – through introduction of the phenomenon of time – is capable of posing many seeming paradoxes. One can only speculate as to when and what stage of our evolutionary pilgrimage this particular mystery will yield its secret.

Summary

It is certainly quite remarkable that galaxy associations, with highly discordant redshifts,* can be made to give very plausible separations in terms of principles inherent in the "New Cosmology." And yet, this is but one example of a host of celestial anomalies which may be resolved upon the basis of a revised concept of radiation propagation and the premise of a gravitational repulsion factor. In particular, we now have before us an explanation for our expanding universe without recourse to a Big Bang cosmology.

Once considered an adequate explanation for the phenomenon of gravitation, Newtonian mechanics is now seen to require modification by no less than three additional equations. By way of recapitulation, it may be stated that this enigmatic force of nature is likely to be conducive to expression as follows:

1) The traditional Newtonian *inverse-square law*.

2) A so-called relativistic *adjustment for motion*. (In reality, this formula depicts the declining efficiency of gravitation with an increase in relative motion.)

3) The prescribed *distance/repulsion* (D/R) formula, in which a portion of exchanged energy is deemed to induce repulsion.

4) A *time-lapse* factor (for which exact calibration is pending), as no physical reaction can take place in zero time.

---

* An informative treatise of discordant redshifts may be found in a book by the eminent astronomer, Halton C. Arp. This valuable work is entitled *Quasars, Redshifts and Controversies* (published by Interstellar Media, 1987; ISBN 0-941325-00-8).

Cosmic Perspective

Noting that the D/R factor can account for the observed expansion of the universe – without recourse to a Big Bang cosmology – it is now evident that this same premise is capable of clarifying a variety of "missing mass" mysteries. In turn, a study of quasar distribution – in light of revised views of radiation propagation – has served to reveal a startling picture of a fateful cosmic edge linked closely to the issue of creation and reincarnation.

Science seems to have a way of progressing in cycles, exhibiting periods of stagnation interspersed with sudden leaps forward in response to new ideas and advanced technology. There is every reason to believe that conditions are favorable to remedy misconceptions that have clouded such fundamental issues as gravitation, propagation of radiation and cosmic origin. Hopefully, in assessing all of the many revolutionary implications of the "New Cosmology," the scientific community will not allow the instinctive prejudice of man to refute ideas simply because they require admission of past error.

But a truly monumental and totally unexpected challenge which now confronts mankind is for astronomers to more accurately determine our proximity to an impenetrable shell of trapped energy so believed to enclose our spherical physical universe – a barrier which threatens to one day extinguish the planet Earth with the swiftness of a lightning bolt! Nor may it be thought that the gauntlet has been cast only to members of this one noble profession. In the years to come, there are crucial roles to be played by both philosophers and physicists alike. Indeed, it may be inferred that the evolving science of cosmology, while currently in its infancy, is destined to become more and more the realm of philosophers.

Not the least intriguing of the challenges confronting generations of future philosophers is that of refining and elaborating upon the inferred universal "time constant." Conceivably, in pursuing such concepts as reincarnation and cosmic fusion, the modern philosopher could become the high priest of scientific research.

(Incidentally, should it ever be wondered whether sufficient energy exists – in the supposed vacuum of space – with which to account for the adduced properties of gravitational attraction and repulsion, one has only to perform a few simple calculations. Upon the basis of proposed theory, it may be stated that there are of the order of some 10^{36} quanta – ranging from "**A**" to below the electron level – per cubic centimeter of volume throughout the expanse of the entire universe! On average, there would be seen to be a flux density of about a trillion trillion trillion minute corpuscles in every 3/8-inch cube of space. In regions populated by stars and galaxies the flux would, of course, be far higher. Hence, it is to be expected that extragalactic radiation will be influenced – in no small manner – during its lengthy passage through swarms of outward flowing quanta.)

Appendices
(PART ONE)

Appendix 1:

Inferred Separations of Galactic Associations

Companion Systems	Distance of Milky Way from Edge, Separation in LY				
	10 MLY	**5 MLY**	**3 MLY**	**2 MLY**	**1 MLY**
NGC 7603	263,000	131,500	79,000	52,600	26,300
AM 2006-295	696,000	348,000	209,000	139,200	69,600
NGC 1232	157,000	78,500	47,000	31,400	15,700
AM 0328-222	577,000	288,500	173,000	115,400	57,700
AM 059-4024	310,000	155,000	93,000	62,000	31,000
AM 2054-221	1,107,000	553,500	332,000	221,400	110,700
Stephan's Quintet:					
NGC 7320 from NGC 7318B	163,000	81,500	49,000	32,600	16,300
NGC 7320 from R = 6,700 km/s Members	197,000	98,500	59,000	39,400	19,700
NGC 7318B from R = 6,700 km/s Members	33,000	16,500	10,000	6,600	3,300

Appendix 2:

Evolution of "M" Stages

Status or "Step"	Status "M" Stage	Quantity Individual Entities	Individual "A" Content	Number of Worlds	Quantity per World
	Prior to Man				
309		2×10^{27}	5×10^{92}	10^{22}	2×10^{5}
310	M^{1}	10^{27}	10^{93}	10^{17}	10^{10}
311	M^{2}	5×10^{26}	2×10^{93}	10^{17}	5×10^{9}
312	M^{3}	2.5×10^{26}	4×10^{93}	10^{17}	2.5×10^{9}
313	M^{4}	1.25×10^{26}	8×10^{93}	10^{17}	1.25×10^{9}
314	M^{5}	6.25×10^{25}	1.6×10^{94}	10^{17}	6.25×10^{8}
315	M^{6}	3.12×10^{25}	3.2×10^{94}	10^{17}	3.12×10^{8}
316	M^{7}	1.56×10^{25}	6.4×10^{94}	10^{17}	1.56×10^{8}
317	M^{8}	7.81×10^{24}	1.28×10^{95}	10^{17}	7.81×10^{7}
318	M^{9}	3.91×10^{24}	2.56×10^{95}	10^{17}	3.91×10^{7}
319	M^{10}	1.95×10^{24}	5.12×10^{95}	10^{17}	1.95×10^{7}
320	M^{11}	9.77×10^{23}	1.02×10^{96}	10^{17}	9.77×10^{6}
321	M^{12}	4.88×10^{23}	2.05×10^{96}	10^{17}	4.88×10^{6}
322	M^{13}	2.44×10^{23}	4.1×10^{96}	10^{17}	2.44×10^{6}
323	M^{14}	1.22×10^{23}	8.19×10^{96}	10^{17}	1.22×10^{6}
324	M^{15}	6.1×10^{22}	1.64×10^{97}	10^{17}	6.1×10^{5}
325	M^{16}	3.05×10^{22}	3.28×10^{97}	10^{17}	3.05×10^{5}
326	M^{17}	1.53×10^{22}	6.55×10^{97}	10^{17}	1.53×10^{5}
327	Stellar Black Hole	7.63×10^{21}	1.31×10^{98}	10^{22}	1

Appendix 3:

Extragalactic Hydrogen Clouds

By reason of their intense luminosity, which is often a hundred times or more than that of a giant galaxy, quasars can be readily seen at great distances. En route to us across the depths of space, this radiation is bound to interact with vast clouds of hydrogen gas which are mandated by the Steady-state model of creation. As a quasar's light passes through such a cloud, the hydrogen will absorb light at very specific wavelengths, with the most notable of these discrete absorptions being the Lyman-alpha line located at a rest wavelength of 1,216 angstroms. But since all extragalactic hydrogen clouds share in the expansion of the universe, it follows that such absorption lines will be shifted to longer wavelengths commensurate with the velocity of recession and distance of these intervening clouds.

Upon analysis of Lyman-alpha lines in the spectra of a sampling of many distant quasars, it was found that the number of hydrogen clouds increases (with higher redshift) at a rate greater than expected from a simple increase of volume with distance. As was the case with an apparent "surplus" of radio sources, this has been misinterpreted as evidence supporting the Big Bang cosmology, since it is deemed that these clouds were far more numerous in the remote past when the universe was only a fraction of its present size and much of the hydrogen had yet to condense into stars and galaxies. Although now seen to have a perfectly logical explanation upon the basis of the "New Cosmology," it becomes of paramount importance to examine Lyman-alpha absorption lines at relatively low redshift. (Should an appreciable number of uncondensed hydrogen clouds be found at modest redshifts, it would only be explicable in terms of the Steady-state principle featuring newly created matter.)

Prior to the Hubble Space Telescope, it was not possible for astronomers to observe Lyman-alpha lines much below $Z = 2.00$, as such radiation is absorbed by the Earth's atmosphere. However, the results are now available and reveal the presence of hydrogen gas clouds inhabiting the regions between nearby superclusters. By way of an example, the ultraviolet spectrum of 3C273 (about 2 BLY away) reveals no less than five major intervening extragalactic hydrogen clouds. Moreover, their spacing is remarkably uniform and is in excellent agreement with separations to be inferred from a Steady-state/Small Bang creation model!*

Clearly, evidence that was first thought to favor the Big Bang concept has once more been disclosed to lend strong support on behalf of the principle of Continuous Creation.

---

* "A Quasar Lights up the Universe," *Astronomy,* September 1991, pps. 42-45.

Appendix 4:

Gamma-Ray Bursts

Support for a distance/repulsion (D/R) factor, as an intrinsic component of gravitation, has emerged somewhat fortuitously from a study of gamma-ray bursts.* Upon examining some 260 gamma-ray bursts, detected by NASA's Compton Gamma Ray Observatory, and plotting the relationship between quantity and intensity, astronomers naturally assumed that their extragalactic origin would assure detection of many more dim flashes than bright ones – just as there is found to exist far more faint, distant stars and galaxies than bright, nearby ones. Although the exact cause of such highly energetic bursts is still a matter of debate, it has been theorized that collisions or implosions of compact massive objects – possibly of the order of neutron star/black hole entities – may be at the root of these violent emissions. (It is conceivable that these brief outbursts could signal a collapse of neutron stars into black holes.) In any event, a deep space origin is clearly indicated by reason of an ubiquitous distribution, since they would otherwise show an overwhelming preference to be located on the plane of our Milky Way.

Instead of finding a steady and progressive increase in the number of fainter bursts, as should be expected with distance, what was revealed was an abrupt slowing of this increase midway between bright and dim bursts. By admission, a most logical explanation has been conceded to be one in which the universe is viewed as expanding at an accelerated rate, rather than being retarded by gravitation, as required by a Big Bang cosmology! This simplistic interpretation of evidence is generally ignored, however, by a preponderance of both astronomers and physicists, who have chosen to cling to an outworn creation model and incomplete laws of gravitation. Unfortunately, acknowledgment of a cosmological constant (D/R factor) continues to be resisted long past the point where unbiased science should have revealed the Big Bang concept to be without foundation.

---

* *Science News*, June 18, 1994.

Appendix 5:

The "Missing Mass" Dilemma

Rather than infer a repulsive side to gravitation, in order to account for excessive orbital motions which so characterize galactic extremities and bound galaxy clusters, astronomers have for some time ascribed this anomaly to the presence of invisible "dark matter." In essence, they believe that it is the gravitational pull of this unseen mass which is responsible for accelerating visible stars and galaxies. In the eyes of many, the most likely candidate for any hypothesized "missing mass" involves very large numbers of hitherto undetected faint red dwarf stars of low mass.

Impressive evidence has recently surfaced which precludes this most favored scenario.* With the refurbished Hubble Space Telescope now capable of detecting faint red stars about a hundred times dimmer than ground-based instruments, it has come as a surprise to Big Bang supporters to find that such low mass stars are far less numerous than what is required by theory – effectively ruling out the possibility that unseen red dwarf stars could contribute significantly to any hidden concentrations of matter!

Oblivious to the idea of a "missing physics" solution, based upon a D/R factor inherent in gravitation (which would thereby eliminate the Big Bang creation model), astronomers continue to grope for an explanation. A noted astrophysicist, John Bacall, comments upon this latest failure as follows: "The dark matter problem remains one of the fundamental puzzles in physics and astronomy."

---

* "Red Dwarfs Can't Explain Dark Matter," *Astronomy,* April 1995, p. 22.

Appendix 6:

Calibration of Hubble's Constant

Efforts to calibrate Hubble's constant (the rate by which the universe is expanding) have been at the frontiers of astronomical research for well over half a century. Today, the consensus of opinion among most astronomers, utilizing a variety of methods, is that the universe has an age/radius of somewhere between 10 and 20 billion years. However, this lower figure is in open conflict with strong evidence that the oldest stars of our Milky Way's globular clusters were formed at least 15 billion years ago. For this reason, an antiquity toward the higher estimate is generally preferred in an attempt to reconcile cosmic age with the firm dictates of stellar evolution.

As it turns out, there is conclusive proof of Big Bang fallacy merely by pausing to examine redshifts within the confines of our own Local Supercluster. It so happens that the average redshift (per MLY of separation) is of the order of 22 km/s. But since members of our supercluster are constrained by gravitational attraction, *this would require the true rate of cosmic expansion to be a great deal higher!* Therein lies an insoluble problem for supporters of all forms of Big Bang cosmology, since admission of higher redshifts (per MLY) only serves to compound an already serious contradiction. (A redshift of 17 km/s would imply an age/radius of about 18 BLY; while 22 km/s gives a figure of 14 BLY; 30 km/s implies 10 BLY; and 40 km/s only 7.5 BLY.)

The problem for Big Bang proponents gets rapidly worse when we assess galactic redshifts in relation to distance traveled since the time of their birth. If we tentatively adopt a figure of 15 BY as being consistent with both Hubble's constant and the ages of our oldest stars, then it may be calculated that a system now 20 MLY distant (with an average velocity of recession of some 330 km/s over the past 15 BY) will have receded about 16.5 MLY since forming out of the interstellar medium. A galaxy currently situated 40 MLY distant will have moved away by not less than 33 MLY; while another at 100 MLY will be adduced to have traveled over 82 MLY in this very same interval. Should we go back in time to their formation we are faced with the rather ludicrous situation of having condensed our entire Local Supercluster, consisting of thousands of galaxies, into a tiny volume of space only 30 to 40 MLY in diameter! Such enormous density would clearly possess so strong a gravitational field that no system could ever manage to escape to greater distances. Upon the basis of a solitary Big Bang explosion there is simply no way in which they could have receded to their presently observed positions – revealing that there is something drastically wrong with the idea that such galactic redshifts depict true motion of recession. This radiation does, in fact, reflect a substantial *time delay* factor imposed by reason of interaction with the inferred D/R quanta and the ubiquitous interstellar medium.

However, while conclusively ruling out the Big Bang model, upon the premise of a Small Bang/Steady-state cosmology, it is logical to expect that a portion of these redshifts will actually be due to residual movement away from the site of our Local Supercluster Small Bang explosion – an event which is presumed to have occurred over 15 billion years ago in the vicinity of M87, at the heart of the dense Virgo Cluster. Clearly, the task before us is to assess the relative importance of these two factors. Although much study will be needed in order to more fully resolve this issue, it is conceivable that a tentative estimate of 50% actual motion (away from M87) and an equal proportion of "time delay" may not be entirely out of line. (If so, it is likely that velocities of roughly 1,000 km/s will have to be subtracted from the redshifts of distant celestial objects to allow for passage of radiation through our own supercluster. In addition, there is also the effect of D/R arcing and the relatively tenuous but extensive extragalactic medium to be considered.)

Thus it will be seen that galactic redshifts cannot be used as a means of establishing Hubble's constant through direct extrapolation. Only by measuring the redshifts of a number of superclusters – over distances of at least a billion light-years and by some totally independent means – may we make a proper determination of this vital cosmic constant. Even then, it will be necessary to consider this "time delay" component of all extragalactic redshifts before such measurements can be deemed truly trustworthy.

Meanwhile, based upon available criteria, it seems reasonable to presume that a figure of roughly 17 km/s (per MLY) may be adopted as being as good as any other in mapping our universe. This would yield a cosmic radius of some 18 BLY – being the point where recession is equal to

APPENDICES

the speed of light relative to the center of the universe. Hubble's constant, as described previously, would then translate to some 12 billion years.

> Please visit the author's Web site for updates of *New Cosmology* research:
> www.webservr.com/science.

Please visit the author's Web site for
updates of New Cosmology research:
www.webserv.com/science.

PART TWO:
Religion and Society

6

The Issue of Biblical Credibility

Inspired Prophets/Priests? – Old Testament Authorship – New Testament Authorship – A Legacy of Unfounded Doctrine – Biblical Symbolism

Tradition frequently leaves invisible mental scars. All too often misconceptions implanted in childhood persist throughout the lifetime of an individual, effectively thwarting the revelations of logic and science. Nowhere is this more true than with the exposure of young minds to erroneous biblical concepts. It is, in fact, little short of incredible that widespread ignorance of this body of literature should still persist in the 20th century, in spite of obvious contradictions with both science and reason.

To those whose religious background would deny such principles as evolution and reincarnation, it is perhaps expedient to remove any vestige of doubt by going straight to the difficulty. For never, in all the annals of history, have so many souls been so misled by the written word of man.

Inspired Prophets/Priests?

Essential to a proper understanding of biblical scripture is the realization that these documents are not the result of any special revelation stemming directly from God and bestowed most suspiciously upon but a few chosen individuals. On the contrary, for while there are indeed instances of inspiration, in a great many respects their opinions must be considered quite inferior to the views of modern man with his superior grasp of science. Nor can it be said that works like Genesis contribute anything useful with regard to the more basic mysteries surrounding the creation of life and the universe. Instead, the Old Testament writings depict a lengthy and often very painful struggle of the Jewish race to develop their theology – embracing the concept of a monotheistic God – in the face of hostile neighbors with multiple deities.

Those occasions where God is supposed to have "spoken" to various select recipients must obviously submit to reason. Clearly, incidents ascribed to the actions and commands of God are merely those which appealed to the so-called prophetic writers. For in their trying endeavor to lead the less spiritually evolved masses further along the road to Utopia, the ancient prophets could evidently see no harm in expressing – frequently by means of symbolism or in the name of God – certain ideas and events which they wished to convey. In an era where the masses were prone to accept authority and were unaccustomed to thinking for themselves, this ploy was generally quite successful.

As the unfounded and highly nebulous dogmas of so many churches reveal today, such an end has unfortunately failed to justify the means. It is an unhappy fact that once an incorrect and provisionary interpretation becomes accepted, tradition will only too likely have the adverse effect of deferring later recognition of the real truth. While this state of affairs may have served a useful purpose with respect to ignorant masses of the past, it is certainly time to admit that the prophets signed God's name to their particular views because it then seemed expedient for them to do so. Most assuredly, we of a more enlightened generation must awaken and treasure our conscience above the frequently restricting influence of tradition. With the passage of time, is it not to be expected that the quality of this conscience will be higher, and that many of our more progressive modern thinkers must have a concept of reality far superior to the views of the ancient scriptural writers?

In order to assess the respective roles of various Hebrew priests in composing the books of the Old Testament, it is perhaps instructive to consider a somewhat nefarious side to their lives. Although they professed to have a pipeline to God, in reality they were in a poor position to expound an enlightening philosophy. With little incentive to engage in open discussion and debate over issues already decided by tradition, a quest for prestige and income soon came to dominate their lives. Perpetrating the colossal hoax of animal sacrifice – through Temple-approved priests only – ostensibly in belief that this act would elicit favor from God, for centuries a high priest and his minions were able to exact considerable tribute from a gullible populace. Throughout the course of much of this archaic literature there was, in fact, a constant power struggle – among rival priestly factions – to secure an exclusive franchise in this lucrative but foolish practice. Along with other clues, this priestly struggle is able to supply a measure of insight into the authorship of certain Old Testament documents.

Old Testament Authorship

Essentially, the Old Testament is a mixture of actual and highly exaggerated history, folklore, superstition and outright myth. Also, as we shall presently see, it contains an ingenious form of *symbolism*. Handed down by word of mouth and scattered writings for centuries, it was not until the great prophetic era extending from about 900 BC to 300 BC that much of this material was finally gathered together, by an assortment of editors, to form a connected series of documents or books.

The first five of these books, comprising the nucleus of the Old Testament and known as the Pentateuch – Genesis, Exodus, Leviticus, Numbers and Deuteronomy – consist of four primary documents: **J**, **E**, **D** and **P**. A brief synopsis of these historical works follows:

J: This Yahweh/Jehovah document is thought by many scholars to be slightly older than **E** (about mid-9th century BC). With an unknown authorship which claimed descent from Aaron, their priesthood was centered in the city of Jerusalem, in the southern kingdom of Judah.

E: Worshipping a deity by the name of Elohim, their priests were reputed to be descendants of Moses. This unidentified group of northern kingdom priest/prophets was likely centered in the city of Shiloh at the time **E** was composed.

D: Comprising the script of Deuteronomy, this document appeared rather suddenly in the year 622 BC, during the reign of King Josiah. Believed to have been written by Jeremiah, and Baruch his scribe, it has much in common with **E** – except that it strongly supported the idea of centralized sacrifice in the Jerusalem Temple. (These same writers are also deemed to have been the chief authors/editors of six other books: Joshua, Judges, I and II Samuel and I and II Kings. The aforementioned six books are often referred to as the Early Prophets.)

P: Known to scholars as the Priestly document, it was evidently written as an alternative to **J** and **E**, with roots extending back prior to the formulation of **D**. The document itself was almost certainly composed by a group of Aaronid priests following the Assyrian invasion of the northern kingdom in 722 BC, very possibly during the reign of King Hezekiah (727 to 699 BC).

P is the largest of the four sources of the Pentateuch, being equal in size to that of the other three combined. It expounds a more cosmic and less personal account of the sevenfold creation narrative and the great flood story. Included are sagas of Abraham, Jacob, Moses and the exodus. It also contains a sizable corpus of priestly law and covers roughly thirty chapters of Exodus, Numbers and the entire text of Leviticus. Both **J** and **E** tell of very similar stories, but with a number of revealing contradictions.

To understand how these four works became interwoven into the early "Torah," it is well to picture certain historical events which played a prime role in this merger. After the death of King Solomon, the Jewish nation was split into two kingdoms in the year 933 BC: King Rehoboam in the southern kingdom of Judah; and King Jeroboam in the northern kingdom of Israel. Two priesthoods emerged with very different ideas as to who was the rightful high priest heir. When the northern kingdom fell to the Assyrians in 722 BC, many Israelites fled to Judah – bringing with them their treasured **E** manuscript. In due course, **J** and **E** were combined for the sake of unity. But the greatest and most difficult task of editing involved the combining of many duplicate (but far from identical) stories of **P** with **JE**. This merger is believed, by many scholars, to have been the work of an Aaronid priest by the name of Ezra, subsequent to his return from a Babylonian exile about the year 458 BC. Written originally as an alternative to **JE**, in a stroke of irony **P** ends up being combined with the very text it was designed to replace. The mystery as to why so many Old Testament stories seemed to be duplicates, and with so many subtle contradictions, was at last clear to biblical scholars once they had separated text into its initial root sources.

A key ingredient in this biblical detective case was the way in which the authors of **J, E, D** and **P** favored certain kings and priestly laws – especially laws relating to the idea of centralized Aaronid sacrifice. For instance, King Solomon and King Hezekiah are depicted as tyrants by **E**; while **P** praises both of these early kings, chiefly because of their support of the Aaronid priesthood. Also to be mentioned, by way of distinguishing authorship, is the fact that while **JE** saw no harm in introducing stories of talking animals and glowing angels in order to impart a message, **P** was opposed to such concepts and made no mention of these obvious fabrications. It is a credit to the diligent work of a number of contemporary biblical scholars that the false notion of a Mosaic authorship of the Pentateuch has finally been laid to complete rest – after persisting for more than two-and-a-half millennia!*

On the whole, it may be concluded that the various Old Testament documents were written by a series of priests with two primary ideas in mind. First, they wished to expound their own personal views and aspirations. Second, they desired to convince an uneducated and naive populace of their superior priesthood status – a trait not exactly absent in this present age.

New Testament Authorship

The very notion that a mere human being could ever be a recipient of an actual manifestation of God can only be described as childish and has all the aspects of a story based upon sheer fantasy and "fairy tale" magic. It simply does not afford a realistic picture of the dynamic cosmos now revealed by modern science. It fails utterly to consider the vast multitudes of other star systems, with their myriad habitable planets, which must exist in a vast celestial universe. Do these other worlds also have a "Christ" such as ours? Do our highly esteemed theologians even know what they are talking about, in their support of such a wild idea – a concept which originated centuries ago and in an era which could conceive of nothing beyond a solitary Earth and a nebulous "heaven"?

In sticking to historical fact, and avoiding the temptation to interpret literally certain biblical narratives that were clearly designed to convey a moral or principle, we may attempt to place the various New Testament writings into proper perspective. Considering that the traditional canon consists of some 27 books, it will be necessary to restrict comment to the main documents comprising this body of literature. (At one time there were

---

* A most informative reference source which may be cited is *Who Wrote the Bible?* by Elliott Friedman (published in 1987 by Summit Books, ISBN 0-671-63161-6).

as many as 50 or more "gospels" in circulation. It was not until after the middle of the 4th century AD that the present canon was decided upon.)

Jesus himself is now believed by historians to have been born between 6 and 4 BC. According to the biblical records, which is all we have to go by, he was crucified during the governorship of Pontius Pilate in Judea (26 to 36 AD). If we accept as fact the mention in the Gospel of Luke (Luke 3:1), where it is stated that Jesus began his preaching in the 15th year of the reign of Tiberius Caesar, a ministry of one year would put his death at about 29 AD. A possible ministry as long as four years would serve to extend his life until around 33 AD. In any event, there is every reason to suppose that his teaching came to an abrupt and unfortunate end not later than this time, and at the hands of Roman soldiers acting under the orders of Pilate.

The oldest written biblical account dealing with Jesus appears to be the various epistles of Paul (dated from 50 to 60 AD) – being the works of a man who had never even seen Jesus and was not converted from the role of a persecutor until some years after Jesus' death. (Although a few scholars hold that I Peter could date back as far as the early 60s AD, it involves the assumption that it was indeed written by the apostle Peter. But since this short epistle imparts little more than encouragement to keep the faith, one must wonder how this relatively trivial document survived while other – presumably more important – works by the same author do not exist. Thus it is likely to be the product of an unknown who was desirous of expressing his views under the name of one more prestigious than himself, and with a date of the order of late 1st or early 2nd century AD.)

The four popular gospels (Matthew, Mark, Luke and John), which are purported to tell the story of Jesus in some sort of chronological format, were all written decades after the crucifixion. The Gospel of Mark (early 70s AD) is widely regarded as being the oldest; while Matthew is probably next, written in the 80s AD. Luke is now thought by some to be dated early 2nd century. The Gospel of John is believed by many scholars to have first appeared on the scene sometime in the 90s AD, long after the apostle by this name must have passed away. Mark was not one of the original disciples, only entering into the picture somewhat later, allegedly as an interpreter of Peter. The work attributed to Matthew is evidently the product of an unknown author who secured much of his information from the Gospel of Mark. (An eyewitness would not have had to depend upon a previous work.) Luke is reputed to have been a companion of Paul and, since neither knew Jesus, would likewise have had to obtain information from an outside source – presumably, from the earlier record of Mark. (There is good reason to suspect that this gospel was written by an unknown who chose to use the name of Luke.) John's version differs in certain respects from the others, being more concerned with the spiritual aspect than any historical Jesus. (It has been determined that another gospel – designated as **Q** by biblical scholars and predating all four canonized gospels – once existed, and that many "sayings of Jesus" were gleaned from this lost manuscript.)

The Acts of the Apostles, ostensibly authored by the same Luke of the third Gospel, has been dated at possibly early 2nd century AD. Yet another major work of the New Testament, the Book of Revelation, also belongs to this same period of time. Using the name of John, it is the work of an unknown scribe. An assortment of letters, dated for the most part well into the 2nd century AD, clearly reflect dogmas of the Jesus-Christ myth, which by now had become firmly established.

In any case, the bottom line is that virtually all of our biblical records of Jesus turn out to have been written by persons who had no firsthand knowledge of their subject! (Indeed, II Peter is conceded, by most authorities, to have an antiquity no earlier than about 150 AD – well over a century after the crucifixion!) Thus, it will be seen that ample time and opportunity existed for a host of new and totally foreign ideas to have arisen before the teachings of Jesus were committed to scripture – including interpolations of many later authors – and subsequently incorporated into church canon. During this interval there is good reason to suspect that such drastic changes in doctrine occurred as to completely alter and rewrite the philosophy taught by its alleged founder – *changes to the extent that Jesus himself would have been astounded and even shocked by the end result!*

A Legacy of Unfounded Doctrine

One immediate and fundamental problem which arises concerns the issue of Christian origins, in which Jesus the man became transformed into a combination Messiah and Redeemer – to say nothing of equating a human being with God Himself! While it must be fully apparent that neither Jesus, nor his close friends and associates, really believed that he was a divine entity, with all the supernatural powers now attributed to him, how are we to explain the seemingly contrary views expressed in the Bible and long since held by Christian churches to be actual fact?

To understand why the New Testament appears to support a Messianic philosophy we must realize that, at the time of Jesus, the Jewish people lived in constant expectation of a Messiah who would come and establish a new world order – a belief which had been fostered by many of the ancient prophets desirous of lending encouragement to the often oppressed Hebrew race. Subsequently, it will be seen that, following the death of Jesus the teacher, in such an atmosphere it would be only a matter of time until advantages were to be had in the preaching of a Messianic gospel. Since past experience with so many Old Testament writings had shown that acceptance of a particular doctrine by the masses was largely dependent upon the prestige of its author, it was most desirable to attribute the highest possible prestige to the master whose works they wished to propagate.

Thus, while the idea of portraying Jesus to be a Messiah may well have been unforeseen before his death, his martyrdom certainly posed a great opportunity to claim a divine origin and thereby obtain much valuable prestige and authority. As to the question of whether or not Jesus deliberately laid down his life for the sake of Principle, we can only surmise. However, inasmuch as his teachings were so opposed to traditional views we must consider that he was not unaware of the dangers involved, and that for the sake of conscience, which must ever lead one to seek and expound the truth – regardless of consequences – he was prepared to risk adverse reaction. If this is so, then we may choose to look upon the crucifixion of Jesus as truly a deliberate act of unselfish love, motivated by conscience, for the benefit and example of all mankind. (On the other hand, as will be discussed in *Chapter 9, Christianity and the Dead Sea Scrolls,* there might well be quite another side to this picture. Perhaps, upon reflection, there could turn out to be a measure of truth to be found in both scenarios.)

In any event, in view of the antiquity of authorship and lack of genuine firsthand information, along with the inevitable "editing" by a series of opinionated writers, it will be seen that the task of incorporating later theological ideas was relatively simple. Indeed, in this respect they were aided by a favorite custom of biblical writers in general who were inclined to word certain stories so that they were capable of a "dual" interpretation. That is to say, if it was to their advantage they liked to combine symbolism with text that was conducive to a literal translation, if readers were willing to give it such credibility. It is, therefore, not surprising that many false ideas were accepted in the course of time and that the status of Jesus is marked by the observed element of confusion.

THE ISSUE OF BIBLICAL CREDIBILITY

Invariably, one of the most important of these false teachings to plague the early church is that of the doctrine of redemption. This belief is somewhat peculiar and apparently found little sympathy among the first Christians of Palestine, as is evidenced by the fact that it did not develop until the new religion had spread out into the pagan world. The Messianic view held by the Jews at the time of Jesus did not center at all upon the idea of a Messiah who would be a Redeemer making atonement for any "original sin" dogma. What they expected was a Messiah who would be more of a teacher and ruler than one who would save by blood atonement! There seems little doubt but what the weird concept of a Redeemer owes much to pagan beliefs of the time.

Mithraism, in particular, seems to have influenced early Christian thought to a very considerable extent. This ancient religion, which originated in Persia as only one of a number that embraced the principle of a Redeemer or Mediator between God and mankind, was firmly established in Rome at least a century before the crucifixion. It flourished until the 4th century AD, when it finally succumbed to the new and highly competitive religion called Christianity. However, this surrender was evidently far from unconditional, since several features of the older religion were clearly absorbed into the basic structure of its successor – *including the very practice of observing Christmas upon the actual birthday of Mithras himself!* Other supposed Christian practices derived from Mithraism include switching the Jewish Sabbath to Sunday, the first day of the week, which had long been dedicated to the Sun and was the chosen holy day of Mithraics. This religion also contributed such ideas as baptism and confirmation, and fostered notions of a mystical salvation from a eucharistic Last Supper.

Until this fateful merger with Mithraism, some centuries after the death of Jesus the teacher, there is every reason to believe that many of the early church members accepted the principle of reincarnation as a viable philosophy. Indeed, since it was not an unpopular belief during the lifetime of Jesus, it is instructive to note that *at no time did he see fit to repudiate it!* On the contrary, such references as "In my Father's house are many mansions" (John 14:2) and "you are of more value than many sparrows" (Matthew 10:31), tend to suggest the principle of reincarnation and (perhaps?) even a Cosmic Pyramid. In any event, history reveals that among the more prominent of the early Christians who embraced the concept were Origen, St. Augustine and St. Francis of Assisi.

If there was one major contributor to the idea that Jesus was a divine incarnation of God it was Paul, an educated Jew from Tarsus. Subsequent to his conversion, this dynamic apostle was sent out of Jerusalem to neighboring countries on a mission to preach the wisdom of Jesus. It was evidently not very long before he saw merit in elaborating upon the works of his idol, in order to more successfully compete with other religions and their semi-god deities. In just a short while, Jesus the teacher was transformed into Jesus-Christ the Son of God. Once this great leap had taken place before one audience it soon became impossible to retract such exaggeration and a new and highly revolutionary religion was born – for better or for worse. (As we shall see, this unfounded claim would be at the root of Paul's later conflict with James, a prominent leader of the early church in Jerusalem.)

Thus it came to pass that many confused souls, desperate for guidance and leadership, fell prey to attractive promises inherent in the new faith called Christianity. In a stroke of irony, with the passage of time even the mighty Roman Empire succumbed to the very religion which it had previously sought to suppress. Spreading from such a strong focal point, it was not long before the concept of a centralized leader was adopted as supreme authority of the Roman Catholic Church: the pope.

Once endowed with the power of an absolute monarch, it was merely a question of time until a host of dogmatic ideas were conceived and thrust upon a populace that was coerced into submitting to priestly authority. Respect for an individual's conscience and freedom of thought were eventually ignored – culminating in acts of outrageous persecution during the infamous inquisitions of the Middle Ages. In compliance with the adage that "absolute power corrupts absolutely," the pinnacle of church folly was attained in the year 1870 when the doctrine of "papal infallibility" was promulgated. In essence, the pope had now become God in matters of faith and morals!

Rebellion against the Roman Catholic Church took a rather bizarre twist as early as the 7th century AD. Originally holding sincere feelings toward Christianity, an Arab by the name of Mohammed became disenchanted by the disunity and bickering among various rival church factions. In particular, certain unpalatable dogma and the spectacle of multiple popes excommunicating each other did not sit well with him, and he felt the need to preach his own version of this religion. By the time of his death (632 AD) the new religion of Islam had gained a firm foothold in the Arab world and was posed to spread to adjoining nations. In yet another example of irony, the Catholic Church had inadvertently spawned a new and highly competitive religion – one which would likewise yield to a temptation to suppress thought contrary to a series of unfounded and convoluted beliefs.

Invariably, there is a vital lesson to be learned from the history of organized religion, with its strong emphasis upon dogma that would forbid criticism and discourage further reflection on the grounds that ecclesiastical authority knows best. But a philosophy based primarily on emotion and blind faith can only serve to deceive the believer. Until we are able to free ourselves from all vestiges of priestly tradition and ponder the universe through the eyes of an open-minded scientist, we will be unlikely to acquire a higher perspective.

Biblical Symbolism

That both Old and New Testament stories are not to be regarded as factual, from the point of historical accuracy, becomes quite evident once we examine the rather unique system of symbolism which pervades these ancient writings. For undoubtedly, the most obvious and most striking characteristic of the Scriptures is to be found in pronounced repetition of particular numbers and cryptographic words or phrases. In fact, when one contemplates the exceedingly lavish use of numerals, it will be noted that practically every such occurrence may be traced to one of a half dozen numbers, or at least to their multiples or to special combinations! Why should this be so? Therein lies a most important and largely obscured facet of biblical literature; for the writers of these documents were concerned with much more than actual history and mere physical events. To them, a chief aim was to convey an intended *moral* or *principle*, quite regardless of whether there was any correlation with reality or whether mythology and fiction had to be resorted to in order to stress the significance of a point.

This they chose to do by devising a system whereby certain names, persons, places, objects, ideas, phrases, happenings and especially *numerals* were introduced for the specific purpose of *representing or symbolizing certain spiritual ideas or principles*. For instance, in connection with the extensively used system of numerical symbolism, the basic numbers 1, 3, 4, 7, 10 and 12 were adopted and given a special meaning in relation to spiritual matters. A brief insight into these subsequently idolized and "sacred" numbers follows:

One:

> The number "one" was held in esteem as a symbol of the recognition of a solitary *Supreme Being* – that there is in reality but one God ruling the universe. An example would be, "could ye not watch with me one hour?" (Matthew 26:40).

Three:

> The figure "three" came to be a rather important symbol which was used to represent the *intimate nature of spiritual matters, of success in realizing or attaining some measure of the truth.* The adoption of this particular numeral may well have been suggested by the overall sum of the conspicuous heavenly bodies – namely, the Sun, the Moon and the stars. A few of the many instances of the usage of the numeral "three" would be the three men who are reported to have appeared at the door of Abraham's tent, Jonah in the whale's belly three days and three nights, the rising of the "Christ" on the third day and the episode of Saul (later renamed Paul) blinded for three days while on the road to Damascus.

Four:

> The symbol "four" was probably derived from the four cardinal points of the compass and perhaps also from the four essential processes of mathematics: addition, subtraction, multiplication and division. Pertaining to spiritual values, this number is used to signify a *sense of direction and spiritual contemplation – the act of seeking the truth* (e.g., the "four winds" mentioned so many times in the Scriptures, the four who are mentioned as setting out for the land of Canaan in the Abraham narrative, the four thousand fed by Jesus, the raising of Lazarus who was dead for four days, etc.).

Seven:

> The number "seven" is accorded a high and prominent position in the Scriptures. Apparently derived from a total of those heavenly bodies known to exhibit motion in ancient times, this figure came to represent the completeness of the heavenly system. Hence the number was used to symbolize a *sense of divine fulfillment and of*

spiritual perfection. (In addition, as will soon be revealed, the number "seven" was also used to convey the spiritual meanings of the seven days of creation.) A few examples of the use of this number would be the seven days of creation, Moses' seven ascents of Mount Sinai, the numerous sevens appearing in Joshua and which concern the falling of the walls of Jericho, the fact that (in the genealogy listed by Luke) Jesus is depicted as the 77th personage, the seven messages, the seven seals, the seven trumpets, the seven vials of Revelation, etc.

Ten:

The adoption of the figure "ten" as a symbol is believed to stem from the sum total of our fingers upon which we depend to perform so many human tasks. Thus the symbol "ten" was selected to depict the actual *application of spiritual thought, of striving to put one's insight of the truth into practice.* Examples would be the Ten Commandments, the parable of the ten virgins, Jacob's solemn statement (Genesis 28:22) that "I will surely give the tenth unto thee," etc.

Twelve:

The symbol "twelve" is evidently derived from the twelve signs of the zodiac, which at one time were held to influence and determine the actions and fate of man. Consequently, the numeral "twelve" was adopted to symbolize a *demonstration of true government, of the authority and universal extent of divine principle* (e.g., the twelve sons of Jacob, the twelve stones carried across the Jordan, the twelve apostles, etc.).

While the above "basic" numerals are by far the most common and are used generally in the senses described, there are a few exceptions which deserve mention. For instance, the numeral "six" is often used as a symbol of mortal and imperfect thought. Another occasion would be the use of a particular numeral pertaining to the message imparted by one of the seven days of creation (e.g., the number "two" could refer to the message conveyed by the second day of creation, "five" to the fifth day, etc.). In addition, there are also times when the symbolic role of these "sacred" numerals is reversed and used in the *opposite* sense! Typical examples of this usage would be the three temptations of Jesus by the devil, Peter's three denials, Judas receiving three times ten or thirty pieces of silver for betraying Jesus, etc.

Once it is recognized that the Scriptures were intended to impart a spiritual lesson or point of view, rather than record mere physical and historical occurrences, they immediately take on a new and higher meaning – the picture at last begins to make sense. That this most conspicuous aspect was ever doubted is quite surprising in view of the remarkably consistent and obvious usage of symbolism of all forms. Much more astonishing is the fact that so many religious leaders still fail to realize, even today, that such stories as the seven "days" of creation, the Adam and Eve narrative and the various "miracles" were devised chiefly for the purpose of illustrating a particular lesson. (Along with motivation of a spiritual nature, which is generally of fundamental importance, it is to be noted that some passages were written with a secondary purpose in mind. That is to say, they were designed in such a manner as to actually lend themselves to a literal interpretation – if simple minds would be willing to give them this credibility.)

In a way we can perhaps understand and partly forgive the actions of the ancient writers, since they must have had a strong incentive to provide the inquisitive masses with some tangible idea of how the world came into being. But this excuse is hardly valid in our present age of enlightenment, and those who would continue to preach a literal interpretation in the face of reason are in reality false prophets doing much to hinder the advance of truth.

7

The Old Testament

The Seven "Days" of Creation – The Adam and Eve Story – Noah and His Ark – Some Early Biblical Narratives – The Saga of Moses – Examples of Supposed "Miracles" – The Book of Esther – The Trials of Job – Some Examples from Daniel – The Story of Jonah – Summary

Armed with insight into the aforementioned system of constructive symbolism, we are now in a position to make sense of otherwise enigmatic biblical stories. In the process, it will be seen that many characters and events have not the slightest basis in fact and were invented purely for the purpose of conveying a specific principle or idea which the writers wished to express.

The Seven "Days" of Creation

Seizing upon a somewhat similar Babylonian creation myth, certain scriptural writers proceeded to develop it as an introduction to the Adam and Eve story which had been composed some time before. When we recall the symbolic meaning attached to the various numerals, it soon becomes apparent why their creation narrative happens to be extended over a period of seven days, since the figure "seven" is a symbol of fulfillment and completeness.

Although designed to give simple minds an explanation of how our physical universe came into being, in reality the seven "days" of creation may be resolved into a unique symbolic presentation of an instructive moral or principle. Depicting a sevenfold process of evolving spiritual thought, these stages may be stated as follows:

1) The first recognition of *conscience,* of an Intelligence or Mind controlling the universe (i.e., the *light* of the first day of creation, since the Sun, Moon and stars have yet to be introduced), is expanded systematically through:

2) A fuller understanding of the virtues of conscience and the sensing that *it is capable of lifting us above the darkness of evil* (i.e., the firmament or *separation* of the second day);

3) The great spiritual uplifting to be had from following conscience must bring a *tangible sense of reality which can ever be relied upon with assurance* (i.e., the dry land of the third day);

4) The realization that the revelations of conscience are the outcome of a divine *System* or *Principle* influencing mortal thought, endowing one with all virtues and powers and capable of leading one in the direction of truth (i.e., the divine system of the celestial universe epitomized by the *Sun, Moon and stars* of the fourth day);

5) A great sense of *exalted thought and abundance of life* as arising from the progressive revelations made by conscience (i.e., the fifth day where the *fowl of the air* are used as a symbol of exalted thought, and the *fish of the sea* symbolize abundance);

6) The vital recognition that man is to have dominion and that he is the *ultimate* earthly creation; what is more, that his soul is essentially of an *immortal nature,* since he is capable of intimate spiritual communion and identification with God (i.e., the sixth day on which man is created in the *"image"* of God); and

7) Finally, the realization of the wondrous potentialities of which man is capable must bring to him a great sense of security, peace and rest, since the attributes of *immortality* and of *communion with God* are all that could be desired by one (i.e., the *completeness, fulfillment and state of rest* epitomized by the seventh day).

An understanding of the true meaning imparted by each of these seven "days" (or stages of spiritual thought) is of importance to a proper interpretation of the Scriptures. Indeed, recognition of these seven messages constitutes a valuable key which resolves a host of otherwise obscure biblical stories.

The Adam and Eve Story

Immediately following the seven days of creation comes a narrative portraying man in quite the opposite sense imparted by the sixth day of creation, and in which a most unethical reason for our imperfect environment would seem to be given. What is the real purpose behind this seemingly contradictory passage? That the Adam and Eve story cannot possibly receive a literal interpretation is now so obvious as to be beyond question, both from the aspect of historical accuracy and from an ethical point of view. To accept a literal translation of this narrative is to deny free will altogether; *it is to cast utter blasphemy upon the Just nature of the spiritual cosmos!*

It is as strange as it is disturbing that religious groups have been able to perpetuate a sadistic hoax for so long. It is strange to think that so many could have accepted an obvious untruth, and most disturbing to find that those who profess to believe in the principle of Christianity could have preached such a repulsive doctrine, and what is more, been so successful in preventing so many from thinking for themselves. To conclude that our imperfect environment is the result of the sin of one man is more than fantastic; *it is positively unthinkable!* On the basis of a literal interpretation we are not the children of a Just and Loving Creator, but the unfortunate victims of a God of spite! To insinuate that we should be made to suffer for the sins of a man whom we have never encountered is just about the most shocking belief imaginable. For if we are responsible for the sins of those who lived before us, then there can certainly be no such thing as free will – life itself becomes a complete farce! What, then, is the true value of this narrative?

As is typical of so many of the early biblical stories, none of the characters portrayed in this particular allegory ever existed in real life, but are fictitious symbols introduced for the purpose of conveying a moral or lesson. In direct contrast to the conclusion, which must be derived from a literal interpretation, the real meaning of this episode was to signify that man's sense of conscience was, in effect, of a distinctly higher nature than that of all other forms of earthly life. In reality, the Adam and Eve story was written for the express purpose of illustrating the principle of *conscience and free will* – not to deny them, as a literal interpretation must imply!

Upon looking over the text of this plot, the evidence in support of a symbolic explanation is so obvious that it is amazing how anyone could have thought otherwise. First of all, is it not of the greatest significance to note that the tree from which Adam and his wife ate, and through which action our imperfect environment is supposed to have originated, happened to be so aptly termed "the tree of the knowledge of good and evil" (Genesis 2:9)? Indeed, is not the symbolism used to illustrate the recognition of conscience perfectly clear when we read, "For God knows that when you eat of it your eyes will be opened, and you will be like God, knowing good and evil" (Genesis 3:5)? And furthermore, what were the immediate consequences of eating from this magical tree? "Then the eyes of both were opened, and they knew that they were naked ... Then the Lord God said, 'Behold, the man has become like one of us, knowing good and evil ...' " (Genesis 3:7,22).

Thus the reason why Adam – *a fictitious symbol representing mortal man* – is caused to suffer as a result of his actions is seen to be quite natural enough, since a greater knowledge of God's Will must necessarily entail exposure to abuse and persecution from his fellow beings who are less spiritually evolved and who do worship material rather than spiritual values. This point, incidentally, is promptly portrayed in the fictitious saga of Cain and Abel – where the lowly physical sense slays one of higher principles – as well as in several other narratives throughout the Scriptures.

Noah and His Ark

Like the symbolism characterizing the seven days of creation and the Adam and Eve story, the extraordinary occurrences reported in so many biblical accounts were similarly evoked for the specific purpose of imparting a spiritual lesson.

For instance, take the episode of Noah and his ark. Who in all honesty can accept this story as an actual historical event? On the basis of a literal interpretation it is clearly fantastic; the story breaks down completely from every conceivable angle. It is not only plainly ridiculous from an historical perspective, but a deliberate deluge of such destructive proportions is wholly inconsistent and unacceptable from an ethical point of view. Invariably, all attempts to interpret this allegory in a literal fashion encounter so many insurmountable difficulties that one would expect this passage to serve as an incentive to bring theologians back to reality – back to the realization that the authors of the Scriptures were concerned primarily with *illustrating principles by means of symbolism!*

Actually, no such person as Noah ever existed. Like the mythical Adam and his descendants, Noah is simply another fictitious character or symbol introduced for the purpose of expressing a spiritual lesson or moral. As a matter of fact, the symbolic aspect of this particular narrative is so conspicuous and so consistent that it would have to be regarded as constituting an allegory, even if the context of the story itself were within the bounds of reality. Is it not somewhat of a coincidence to read that Noah's father, Lamech, happened to live to the rather strange and unusual age of *"seven hundred and seventy-seven years"* (Genesis 5:31)? Is it not also significant to further read that Noah had *three* sons (Ham, Shem and Japheth); that the length of the ark was *three hundred* cubits (the multiplied sense of the "ten" and "three"), the breadth of it *fifty* cubits (the meaning of the fifth day of creation combined with that of the numeral "ten") and the height of it *thirty* (or "three" times "ten") cubits (Genesis 6:15); that the ark should contain *three* stories; that Noah was *six hundred* years old when he entered the ark (the sense of the sixth day of creation combined with that of "ten"); that it rained *forty* days and *forty* nights (the symbolical meanings of the numerals "four" and "ten" interwoven together); that the ark rested in the *seventh* month, on the *seventeenth* day ("seven" plus "ten") of the month, upon the mountains of Ararat and that in the *tenth* month, on the *first* day of the month, were the tops of the mountains seen (Genesis 8:4-5); and that Noah sent out a *dove* (a symbol of purity and peace) *three* times, at intervals of *seven* days? Indeed, when we also consider that throughout biblical literature the *ark* is held to represent a high and inspired mode of thought, that the *pitch* stressed in its construction is a symbol of atonement or "oneness" with God, and that even the very word "Noah" itself means "rest" or "comfort," does it not become apparent that the whole affair is nothing more than a symbolic presentation of some cleverly conceived moral or principle?

The overall message to be drawn from this spiritually significant allegory of Noah and his ark may be defined essentially as follows: Whenever we find ourselves engulfed by the floods of materialism and despair, we may find safety and solace by casting our eyes in the direction of God; in response to our efforts He will lift us above the ravaging effects of mortal actions and experiences.

Some Early Biblical Narratives

Bearing in mind that the ancient prophetic writers were far more interested in matters of spiritual importance than they were in mere historical facts, let us consider briefly the significance of certain early biblical characters and narratives.

While the purely mythical characters of the Old Testament do give way eventually to real personalities (even if somewhat exaggerated and idealized), there is a lack of tangible proof with regard to the lives of such individuals as *Abraham, Isaac, Jacob* and *Joseph*. However, once it is realized that these personalities were introduced into the Scriptures for the purpose of illustrating certain morals, it will be seen that it makes little difference whether they are fictitious or not, since the principle behind an event is of far greater importance than a mere physical and transitory event.

Abraham

Abraham himself is introduced into the Scriptures for the purpose of portraying a more advanced understanding of man's true relationship with the Father. As opposed to a deep-thinking Abraham, we have the *Lot* state of thought which might be said to represent the worldly and somewhat indifferent attitude toward God which hesitates to expend the necessary effort to evolve spiritually. One other participant in this narrative is Abraham's wife, *Sarah,* who is injected as a symbol of the necessity to keep one's thoughts "wedded" to matters of a spiritual nature.

The actual story gets underway with the recording of *four* people setting out for the land of Canaan. Understanding the significance of the numeral "four" and the subsequent command to "Go from your country and your kindred and your father's house to the land that I will show you" (Genesis 12:1), can we not see that here we are dealing, not with an expedition to some foreign country, but *with an expedition setting out in search of Truth?* Have we not heard, at one time or another, this same call to "Go from your country" – to abandon the enslaving ties of all self-centered and earthly aspirations (including the utterly foolish sense of nationalism) and strike out bravely in search of the freedom and true glory to be found in the Truth? What a great mistake it is to attempt to reduce this allegory to but a mere historical and physical event and thereby miss the lesson of God's constant urging (i.e., the revelations of conscience) to "Go from your country ... to the land that I will show you!"

On the rough and lengthy road to Truth it is recorded that Abraham went "down to Egypt to sojourn there" (Genesis 12:10) and that he promptly got into trouble through falsely stating that Sarah was only his sister. Recalling the symbolic meaning attached to the name of "Sarah," along with knowledge that "Egypt" is often used as a symbol of unenlightened thought, one is enabled to arrive at the following interpretation: In the face of trials and tribulations, and when adherence to our "Sarah" (or obligation to seek the Truth) might bring us into disfavor, are we not all tempted to disguise any unorthodox views and make believe that we are not so closely attached to such beliefs as we really are – that it is not our "wife," but only our "sister"? However, Abraham's sense of reality was too high to permit him to stay long in Egypt, and it is reported that he "went up from Egypt" and "called on the name of the Lord" (Genesis 13:1,4).

As is bound to be the case when an individual's perspective has increased to a point somewhat in excess of those around him, and if remaining together would only hinder the journey to Truth, then a parting of the ways must inevitably take place. Knowing full well that Lot would choose the way of spiritual stagnation, Abraham felt safe in offering Lot the choice of turning to either the right or to the left. Consequently, while Lot chose the easy way – the "Jordan valley" that was "well watered" – and eventually had his very beliefs and foundation destroyed under him (the symbolic destruction of Sodom and Gomorrah), Abraham continued along the rugged upward path to the "land of Truth." Our obligation to seek the Truth (i.e., the command to look in all *four* directions), and the increased clarity to be had once we divorce ourselves from adverse influences (or the Lot state of mind) is well illustrated by the following: "The Lord said to Abram, after Lot had separated from him, 'Lift up your eyes, and look from the place where you are, northward and southward and eastward and westward; for all the land which you see I will give to you and your descendants for ever' " (Genesis: 13:14-15).

And yet, spiritually minded as one may be, there are frequently lapses or temporary periods of regression. This necessity for man always to struggle against his own slothful nature is stressed in the episode of Abraham forsaking Sarah and taking a mere maidservant (*Hagar*) to wife, the result of which union was *Ishmael*, who symbolizes the everyday physical aspect of mankind. Finally, Abraham's concept of spiritual reality rose sufficiently to make him understand that the Ishmael sense of things must be cast loose, regardless of how much we might wish to retain it. That stories such as this are allegories, and were never meant to receive a literal interpretation, is further borne out by the significant reference of Paul to this particular saga in Galatians 4:22-26, in which he commented on this narrative: "Now this is an allegory!"

Isaac

Although obliged to omit many other interesting incidents relating to Abraham, there is perhaps one more regarding the introduction of the symbol *"Isaac"* which should be mentioned. First of all, it is stated that *three* men of the Lord appeared at the door of Abraham's tent and told him that a true son or heir would be given unto him through Sarah. Recalling the meaning of the numeral "three," along with the knowledge that Abraham represented an advanced sense of man's true relationship with the Father, this allegory might be translated as follows: Insight into spiritual reality informed Abraham that his more advanced state of thought could never perish, but must live on and multiply in the minds of others.

Consequently, with the death of Abraham, Isaac – his promised son – now inherited the role of an advanced spiritual thinker pressing ever onward in search of the Truth. Thus we are enabled to grasp the real significance behind Abraham's request that Isaac never return to the country which he himself had left, since one in genuine search of the Truth can never turn back but must continue to forge ahead. (Witness the symbolic episode of Lot's wife who was turned to "salt" when she looked back.) When we recall how Sarah was used as a symbol to Abraham, it is not surprising that *Rebekah* became a similar symbol to Isaac. Further, it is interesting to note that "Isaac was *forty* years old when he took to wife Rebekah" (Genesis 25:20).

Jacob

Just as in the Abraham saga where an advanced state of thought is contrasted with a Lot state of thought, so in the *Jacob* narrative we have a similar parallel with a "Jacob" sense (a sort of Abraham and Isaac) and an *"Esau"* sense (so typical of the average individual). The first significant incident regarding these two contrasting figures lies in Esau's reported selling of his birthright to Jacob for mere physical food. What does this selling of one's birthright signify? It signifies that the Esau state of thought, when put to a test, will be only too likely to succumb and relinquish its "birthright" – that of spiritual insight and potential dominion over all things – for some physical gratification. How frequently do we all tend to loosen our grip upon our own "birthright" for the sake of obtaining advantages which are of but a temporary and material nature!

Shortly after this selling of the birthright and the subsequent blessing of Jacob by his father Isaac, it is reported that the jealous wrath of Esau forced Jacob to leave home. (Is it not characteristic of mortal man to become jealous of another's success – especially when one starting life in less favorable circumstances is observed to pass us and gain possession of something desirable which could have been ours had we been but willing to expend the necessary effort?) Jacob evidently deemed it advisable to flee the wrath of Esau because he felt that his perspective was not yet sufficient to deal with really pressing tribulations. He needed time to construct a philosophy capable of overcoming the influence of an Esau state of mind.

Accordingly, after Jacob fled from Esau it is recorded that he spent the night at a place called Bethel, which means "the house of God." Moreover, it is stated that after "Taking one of the stones of the place, he put it under his head and lay down in that place to sleep" (Genesis 28:11). Since throughout the Scriptures the word "stone" is used as a symbol of the *appearance of Truth,* it follows that Jacob was indeed earnest in his quest. The success of this search is stated in the very next verse: "And he dreamed that there was a ladder set up on the earth, and the top of it reached to heaven; and behold, the angels of God were ascending and descending on it!" (Genesis 28:12). The ladder reaching up to heaven symbolizes Jacob's understanding of the ordered process of spiritual advancement (i.e., the ascending scale of the seven days or stages of creation); the angels ascending and descending are symbolic of seeking (the ascending angels represent ascending thought) and finding some measure of Truth (the descending angels symbolize the "coming down" or entering of conscience – the "Voice of God" – into one's mind). The fact that "Jacob rose early in the morning, and he took the stone which he had put under his head and set it up for a pillar and poured oil on the top of it" (Genesis 28:18) signifies that Jacob decided to consecrate his life to the search for Truth, since "oil" is used frequently as a symbol of *consecration.* Also of interest is the statement that "I will give the *tenth* to thee" (Genesis 28:22), for interpreted it means that Jacob planned to *put his increased knowledge into practice.*

Just as in the case of Abraham and Isaac, Jacob also found it desirable to keep his thoughts wedded to spiritual matters. Hence it is reported that Jacob said unto *Laban,* "I will serve you *seven* years for your younger daughter Rachel" (Genesis 29:18). But after serving the seven years (or more precisely, attempting to use the "sevenfold path to Truth"), Jacob found himself in possession of *Leah,* the older daughter. While perhaps not too poor of a companion, she was still short of his ideal of Rachel. However, Jacob was not one to give up so soon, and by promising to serve Laban yet another seven years he finally won his "ideal," or higher sense of values.

On the way back at last to face a climax with his brother Esau, Jacob was plagued with a struggle inasmuch as he began to fear that his perspective might still be inadequate to cope with the Esau state of mind. Thus at the brook of Jabbok we find Jacob engaged in a "wrestling bout" with himself, the outcome of which was assurance that his superior vision would see him safely through his darkest hours. This assurance is soon illustrated in the changing of Jacob's name to *Israel;* for where "Jacob" meant only "contender," the name "Israel" is symbolic of a "prince of God." (At a similar crossroad in life Abram had previously been renamed Abraham and Sarai, his wife, was renamed Sarah.) The validity of such an opinion was borne out when Jacob and Esau finally met, for while their views remained opposed they abided in peace, physically speaking, and Jacob had naught to fear.

Before concluding the Jacob narrative it is well to mention the significance of Jacob's *twelve* sons. Like all that has come before, this "twelve" is not to be taken literally, but rather is intended to symbolize the universal scope and authority of divine Principle and true Government. The "twelve" sons in this instance infer that the advanced Jacob state of mind must multiply and exhibit authority. (The twelve sons of Jacob are supposed by tradition to have formed the twelve Tribes of Israel.)

Joseph

Of the twelve "sons" allegedly fathered by Jacob, it is somewhat instructive to note that only *Joseph* and *Benjamin* are actually born of Jacob's "ideal" Rachel. In conformity with earlier examples, Jacob's blessing of Joseph meant that this son now inherited the symbolic role of portraying an advanced state of spiritual thought. (The "coat of many colors" which Jacob gave to Joseph represents receiving insight into the essence of the meaning of the seven days of creation.) With the birth of the high Joseph state of spiritual vision, it is stated that Rachel desired for the Lord to "add to me another son" (Genesis 30:24). What is meant by this strange perception that "another son" must be added? The answer is to be found in the *ability to give up the mortal and worldly sense of things* which the "Benjamin" sense symbolizes; for a higher perspective is of little avail unless one is able to really accept – and live – a superior philosophy of life, with no regard for personal cost. Hence it is reported that Benjamin was brought forth through sorrow and deadly anguish; for it is indeed painful to give up many things of this world which we have come to cherish so highly.

As Abraham and Jacob were matched symbolically against opposing and inferior states of mind (the Lot and Esau mentality), so Joseph had to struggle against his less spiritually evolved brethren. The result was that his brethren, being unable to tolerate a superior and more demanding state of thought, stripped him of his coat of many colors and sold him into Egypt. (In their foolish endeavor to thrust a literal interpretation of the Scriptures upon us, have not our theologians of today been guilty of stripping a "coat of many colors" – the light of Truth – from us? In effect, would they not sell us into the same state of darkness as is symbolized by Egypt?) However, as much as one may attempt to suppress spiritual thought and subject us to trials and tribulations, if our philosophy is built upon a firm foundation, then we cannot help but rise above the situation. This is exactly what Joseph managed to do; for instead of remaining in meek submission he rose to such great prominence that even his brethren had to come to him on bending knee.

The incident of Joseph translating the Pharaoh's two dreams (*seven* thin cows consuming *seven* fat cows, and the *seven* thin ears of corn devouring *seven* fat ears) is perhaps deserving of particular comment. Joseph's interpretation that there would be seven years of plenty to be followed by seven years of famine does not refer to matters of a mere physical nature. On the contrary, the moral which the writers of this narrative were attempting to stress is that, as filling as an understanding may be of the spiritual meaning behind the seven days of creation (the seven fat cows and seven fat ears), unless we take care to preserve this light, the influence of our materialistic environment (symbolized by the seven thin cows and seven thin ears – depicting a "counterfeit" sense of the seven days of creation) will destroy our precious light. As Joseph advised the Pharaoh, the best safeguard is to make full use of our opportunities (the seven fruitful years) to build up our storehouses with harvest (spiritual insight), so that in the face of famine or tribulation (the seven lean years) we will have substance to see us through.

That the entire story is simply another instructive allegory is clearly demonstrated when the starving Egyptians came to Joseph for food *three* times. The first time Joseph took all their money, the second time he took all their cattle, and finally, the third time, he took not only their lands but also their bodies (Genesis 47:14-21). Interpreted literally, one can hardly escape the conclusion that here indeed is a profiteer to end all profiteers. But this is not so, for the "starving" Egyptians were not people suffering from any physical hunger, but rather they symbolize a humanity beset by a *spiritual* hunger; as the food which Joseph gave them was spiritual insight. The fact that Joseph took all their worldly possessions from them (including jurisdiction over their physical bodies) symbolizes that only the spiritual is of real importance, since that is all that was left to them!

During this famine it is recorded that while Jacob sent ten sons to buy some corn in Egypt, he also declined to send the all-important and symbolic "Benjamin" – namely, the ability to give up the worldly sense of things. But Joseph realized that, valuable as may be an attempt to apply a measure of spiritual insight to human experience, any such action is quite incomplete without the ability to divorce oneself from materialism and worldly tradition. Hence they did not receive sufficient food until Benjamin was brought before Joseph and all twelve sons were finally united (i.e., the "ten" representing the application of spiritual thought to the human problem must eventually rise to the "twelve," symbolizing recognition and acceptance of divine Principle).

The Saga of Moses

Moses is perhaps the first Old Testament character for whom any grounds can be said to exist in support of a real personage, although it is fully apparent that the story of his life has been altered so as to express the symbolic policies and aspirations of subsequent Hebrew prophets. Indeed, the biblical record of Moses is so saturated with symbolism that a literal interpretation of many of the reported events is obviously out of the question. Recalling that present accounts of Moses were not penned until many centuries after his death, and that at the time of the prophetic writers he

The Old Testament

had come to be idolized by the Hebrews, it will be seen that there prevailed a marvelous opportunity to attribute to Moses any particular document which later editors were desirous of having accepted. Just what views were actually those of Moses and which must be ascribed to a succession of religious prophets becomes a matter of conjecture, although judging by the extensive use of the sacred numerals, the influence of the Priestly era writers alone is most conspicuous.

Conceding the need to restrict comment to little more than a brief outline of the rich symbolism enveloping the life of Moses, a few of the more noteworthy examples are as follows:

The Early Life of Moses

The circumstances surrounding the birth of Moses are somewhat similar to those connected with the birth of the babe Jesus, in the sense that both Pharaohs and Herods alike seem to have a most uncanny ability to predict *the appearance of light*. In both instances the reported slayings of male Hebrew children are not to be taken as a literal fact, but instead symbolize that our imperfect world is only too inclined to suppress or slay anyone seeking to reveal the Truth. The tale of the babe Moses being hid *three* months (symbolizing insight into spiritual matters) and then placed into an *ark* (symbolic of a high and inspired mode of thought) daubed with *pitch* (a symbol of harmony with God) from whence he was found and raised by none other than the Pharaoh's daughter (a symbol of love amidst hate) is simply an allegory stating that while there may be an abundance of darkness and untruth in the hearts of many, the Word of God will never perish but will take root and grow in the hearts of others.

The following story of how "when Moses had grown up" he slew an Egyptian for beating a Hebrew (Exodus 2:11-12), symbolizes the dark "Egyptian" state of mind succumbing to a more advanced philosophy. The fact that Moses elected to flee from the face of Pharaoh is to be attributed to the realization that, though higher than the state of mind he had "slain," his outlook was not yet sufficient to see him through any severe trial. Hence it is recorded that after fleeing he was met at a well by *seven* daughters of the priest of Midian (Exodus 2:16). This meeting of the *seven* daughters symbolizes the appearance to Moses of insight into the treasured sevenfold process of spiritual advancement, as epitomized by the seven days of creation.

The Episode of the "Burning Bush"

In the third chapter of Exodus we encounter a most interesting example of symbolism. It is recorded that after coming to Horeb, the "mountain of God" (a *mountain* of some sort is frequently used as an expression of *exalted thought*), Moses witnessed the amazing sight of a bush burning with fire, "yet it was not consumed" (Exodus 3:2). Translated into plain language this incident informs us that, regardless of how vivid may be the ravaging effects of imperfect mortal minds (symbolized by the fire), it will never succeed in overcoming or consuming spiritual values (represented by the bush).

The reaction of Moses to this wondrous sight constitutes a rather instructive moral: "And Moses said, I will turn aside and see this great sight, why the bush is not burnt" (Exodus 3:3). If, like Moses, we too could be set thinking and inquiring earnestly into matters of real importance, how quickly would present darkness, fear and injustice give way to the light and comfort of Truth! Only when men stop and attempt to ascertain *why* (in spite of our inherent nature tempting us to think exclusively of our own personal prosperity, there happens to be an unquenchable voice within us ever urging us to a sense of universal love) will the world know real peace, prosperity and genuine happiness.

The Three "Signs" Given to Moses

After the "burning bush" incident, which so inspired Moses to spiritual contemplation, it is reported that the task of spreading the resulting philosophy first appeared too formidable. In answer to his fears and doubts Moses was then given *three* "signs" (see Exodus, chapter 4), or in simple language an increased insight into the powers and nature of God. This deeper perception of spiritual reality was to impress upon him some measure of the absolute mastery of Truth – or of identification with God – over such well-known mortal experiences as evil, disease and death. These three "signs" are as follows:

First sign:

> The casting down of a rod (a rod of mortal thought), which "became a serpent," symbolizes the act of exposing our inherent and imperfect nature. The subsequent action of Moses following the advice of God (i.e., conscience) and grasping the serpent by

the tail, thereby causing it to turn into a rod (in this case a rod of firm identification with God), signifies that obeying one's conscience will annihilate the false power of evil and endow us with genuine authority and dominion.

Second sign:

This second sign, involving a leprous hand, concerns faith in the ability of an increased spiritual insight to overcome the injustice of physical disabilities. Just as God may permit physical misfortunes, so these misfortunes will be seen to be of but a temporary nature, since God is both capable and desirous of overcoming such injustices upon a long-term basis.

Third sign:

The third sign, involving the pouring of water (a symbol of the Word of God) upon the dry land (a parched and lifeless mortal concept), resulted in the water turning to blood (this time a sign of life) and signifies that acceptance of the Voice of God will inject life into even the most hopeless cases.

The Egyptian "Plagues"

Like so many other suspicious events in the Scriptures, the report of Moses supposedly instigating a series of highly devastating plagues against the Egyptian people (Exodus, chapters 7 through 11) is not to receive a literal interpretation. Indeed, those who would entertain such a view are actually guilty of the same blasphemy as are holders of the most horrifying doctrine of original sin! Why should an entire race be singled out for punishment just because one man (the Pharaoh) refused to "let my people go?" Clearly, no individual of any intelligence could accept a literal translation of so obvious an allegory!

Actually, the Pharaoh is used as a symbol of imperfect mortal concepts which tend to hold us in captivity and prevent us from reaching the promised land of Truth. The fact that Moses is reported to have asked the Pharaoh *seven* times to "let my people go" is most significant, for it symbolizes that the sevenfold process of spiritual advancement was being applied to this false sense of authority. That the Pharaoh required physical proof as to the necessity to let Moses' people go is not at all surprising, since we ourselves can frequently conceive of nothing which does not have a physical motive. However, in spite of many difficulties, spiritual perseverance must triumph in the end – it is recorded that Pharaoh finally consented to their freedom.

With regard to the various "plagues" themselves, there is every reason to believe that they were used by the prophetic writers to symbolize the inevitable destruction of certain false pagan beliefs which were considered a threat to Hebrew doctrine. Although our modern Bible lists ten plagues, there is evidently confusion as to the nature and number of plagues intended among the early manuscripts from which this story was derived (i.e., the Jahweh, Elohistic and Priestly documents). In any event, taking as examples two plagues which are characteristic of all three documents, we may deduce the following interpretations:

The Nile turned to blood:
> In early biblical times the Nile was actually worshiped by inhabitants of Egypt as the very creative center of life – quite the opposite of the true Intelligence we call God, which is the sole source of Life and alone worthy of worship.

The "slaying" of the first-born:
> This plague, involving the slaying of "all the first-born in the land of Egypt" (Exodus 11:5), refers not to a horrifying slaughter of the first-born of people, but rather of the first-born beliefs held by imperfect man. Since the "first-born" beliefs of mankind can only be described as "Egyptian" or pagan in nature, it will be seen that in order to advance spiritually, it is most essential that our own "first-born" be slain also.

Although having touched but lightly upon the wealth of ideas and morals symbolized by such episodes, it should nevertheless be possible for one to go ahead on personal initiative and apply these same basic principles of symbolism to other allegorical writings. For instance, continuing the saga of Moses, it will be found that the celebration of the Passover, the *seven* ascents of Mount Sinai (in which he received *Ten* Commandments) and even the instructions regarding the building of the Tabernacle all convey messages of a spiritual nature. Moreover, the fact that Moses was

120 years old when he died (Deuteronomy 34:7) is also quite interesting, inasmuch as the symbols *one, ten* and the sense of the *second* day of creation may be discerned. Interpreted, it signifies that Moses recognized one God and sought to impress this principle upon his people, and that at the time of his death he had, in some measure, succeeded in separating good from evil – he had in truth led them out of the land of "Egypt," out of darkest thought.

Examples of Supposed "Miracles"

There was once a time when all the reported "miracles" of the Bible were accepted without question as being authentic historical events. But as the moral of the "burning bush" illustrates, spiritual progress can never be consumed or destroyed. In spite of the adverse influence of unfounded tradition, it seems likely that this present generation will usher in a new era of progressive thought.

At first it may come as somewhat of a shock to many sheltered religious souls to learn that all of the alleged "miracles" are really symbolic; but this is an age of rapidly dawning consciousness in which such views must submit to reason. In fact, the very nature of any miracle – or divine intervention – strikes at the ethical nature of the cosmos; for it involves what may only be described as *favoritism!* Why should any particular person or persons be singled out, in a truly special manner, to experience some extraordinary occurrence which is against the very laws of nature? Is it not clear that an outstanding feature of the cosmos lies in its consistency – that in order to be just it cannot possibly contradict itself?

The Crossing of the Red Sea

Take for example, the story of the "miraculous" crossing of the Red Sea as described in Exodus, chapter 14. To assume that God would intervene directly and in such dramatic fashion to save one nation from its enemies, without showing the same favor to others faced with comparable predicaments, is quite contrary to all sane reasoning: God plays no favorites! In reality, the reported crossing of the Red Sea does not refer to a mere physical passage through a physical medium, but rather symbolizes a *spiritual passage through a spiritual medium!* It symbolizes the passage of man from the darkness and fear of the "Egyptian" state of thought to a superior and more concrete state of mind (i.e., the "dry land" state of the third day of creation). As a matter of fact, there are three distinct stages of thought involved in this allegory which correspond perfectly to the principles expounded in the first three days of creation:

1) "Then Moses stretched out his hand over the sea; and the Lord drove the sea back by a strong east wind all night" (Exodus 14:21)... "and the Spirit of God was moving over the face of the waters" (Genesis 1:2);

2) "And the waters were divided" (Exodus 14:21)... "Let there be a firmament in the midst of the waters, and let it separate the waters from the waters" (Genesis 1:6); and

3) "And the people of Israel went into the midst of the sea on dry ground" (Exodus 14:22)... "and let the dry land appear" (Genesis 1:9)

In the course of our earthly lives we are all faced with the challenge of having to pass over some similar obstacle – of having to improve and expand our perspective in order to escape the effects of materialism and evil. Applied to everyday experiences, the overall moral of this instructive allegory may be stated as follows: Whenever confronted with fears and doubts, and with difficulties which seem to us to be insurmountable, if we will but press onward, ever trusting in the light of conscience to guide us, we shall soon find ourselves to be situated upon dry land; our fears and doubts will eventually be washed away (just as the "Egyptian" state of mind was "drowned"), and we will have succeeded in crossing our own "Red Sea."

The Crossing of the Jordan

The story of Joshua leading the Israelites across the River Jordan is to be considered a sort of sequel to the earlier crossing of the Red Sea and symbolizes a further advance in spiritual understanding. Unlike the previous crossing, which was made of necessity and tinged with an element of fear, the crossing of the Jordan is characterized by a sense of authority, power and dominion stemming from increased insight into the Truth (the *twelve* men carrying the *twelve* stones). This increased spiritual vision is well illustrated in the statement that *seven* tribes (symbolizing a false or opposite sense of the seven days of creation) would be driven out before them: "And

Joshua said, 'Hereby you shall know that the living God is among you, and that he will without fail drive out from before you the Canaanites, the Hittites, the Hivites, the Perizzites, the Girgashites, the Amorites, and the Jebusites' " (Joshua 3:10). The fact that the *ark* of the covenant was carried before them signifies exalted thought and an earnest search for Truth. Also of great interest is the statement that "The people came up out of the Jordan on the *tenth* day of the *first* month" (Joshua 4:19), as it symbolizes the recognition of One Supreme Cause and of the duty to apply such insight to matters of human experience.

The Falling of the Walls of Jericho

Here we come to an allegory permeated so conspicuously with the sacred or symbolic numerals (particularly the number "seven") that it is positively amazing how anyone could read it without being forced to abandon all notions of a literal interpretation. First of all, it is significant to note that *"forty thousand* ready armed for war passed over before the Lord for battle, to the plains of Jericho" (Joshua 4:13). Invariably, the authors of this story were attempting to stress their quest for Truth, and to put their presumably superior knowledge into practice. The statement that *"seven* priests shall bear *seven* trumpets of rams' horns before the ark; and on the *seventh* day you shall march around the city *seven* times, the priests blowing the trumpets" (Joshua 6:4) signifies application of the sevenfold process of spiritual advancement to the problem at hand. The subsequent fall of the walls of Jericho does not constitute a miracle involving mere physical walls at all, but rather symbolizes the fall of much more important "walls of ignorance" which would prevent us from realizing the Truth. What was destroyed was not physical structures and actual living inhabitants, *but a false sense of reality!*

The Episodes of Samson

Although perhaps inspired originally by the life of an actual personality, the "fairy tale" exploits of Samson (Judges, chapters 14 through 16) most definitely constitute another allegory intended to express a moral or principle by means of symbolism. For instance, who could be so naive as to really believe that the *seven* locks (again the "seven") upon Samson's head were responsible for superhuman physical strength and that, in rather magical fashion, the simple act of cutting off these locks of hair resulted in such an abrupt and mysterious loss of his strength? Indeed, even a cursory glance into the text of Samson's adventures is sufficient to reveal its symbolic nature.

For instance, the riddle given by Samson unto *thirty* companions, and which was to be answered in *seven* days, is introduced to depict the instructive system of symbolism (especially the *seven* days of creation) pervading the Scriptures. The *"thirty* linen garments and *thirty* festal garments" to be given as a reward (Judges 14:12) for interpreting this riddle has a logical translation: An understanding of biblical symbolism will endow one with insight into the Truth (i.e., the "three"), giving a recipient an opportunity to put such virtue to good account (i.e., the "ten"). The fact that they could not answer the riddle by themselves signifies that they did not possess a genuine insight into spiritual matters. Thus it is reported that *thirty* Philistines were "slain" as a result of this deception (i.e., one applying the opposite of insight into the Truth is very much in danger of being "slain," since spiritual death is the only true death).

The "wife" which Samson is supposed to have desired, and which the Philistines (a symbol of mortal thought) would prevent him from obtaining, is really a symbol of something of a high nature to be cherished – an increased perspective. The crops which Samson was responsible for destroying symbolize the Philistine's imperfect minds; while the *three hundred* foxes which were involved in the burning of the grain inform us that when insight into the Truth is applied it will have a vivid effect upon all who hold just such an imperfect philosophy. The reprisal by the Philistines, in which it is stated that Samson's "wife" was burnt with fire, symbolizes that when an understanding of spiritual reality is given to those mortal minds incapable of absorbing such virtue, they will invariably seek to destroy the very source of this enlightenment. (It was through his desired "wife" that they were able to interpret the riddle.)

The *"three thousand* men of Judah," who delivered Samson into the hands of the Philistines because they were afraid of retaliation as a result of Samson's actions, may be taken to symbolize that imperfect state of mind which would succumb to worldly matters and surrender Principle for physical well-being. Samson's reply in slaying a *thousand* men of the Philistines with the "jawbone of an ass" is seen to symbolize recognition of but one God, and of the fatal mistake in opposing one with such a concept.

The story of Samson and Delilah itself centers around attempts by the Philistines to steal the secret of Samson's power. The *"eleven hundred* pieces of silver," which each of the Philistine lords offered for an explanation of Samson's unusual strength, symbolizes that they failed to recognize one God (i.e., the "one" aspect) and were living a false philosophy (i.e., the "ten" aspect). The *three* incorrect explanations given by Samson convey the message that mortal thought can only expect to receive a false perspective. However, if one is really desirous of obtaining knowledge, it will be given – even if not understood (and the *fourth* time it was). That the Philistines, when they learned the secret lay in Samson's *seven* locks (the *seven* locks are a symbol of the sevenfold process of spiritual advancement), promptly went ahead and persecuted him (they cut off his locks and mistreated him severely), symbolizes what mortal thought forever wants to do to those with a higher perspective. But while the actions of man might serve to blind a person physically, they cannot suppress the light and strength of a truly superior philosophy – the *seven* locks grew back again. The moral follows that in the end those possessing a false perspective will invariably be destroyed at the hands of one blessed with a higher philosophy. Hence it is reported that *three thousand* Philistines on the "roof" were destroyed as a result of a separation of "two middle pillars" (the "three" is used in a negative sense; while the "two" expresses the meaning of the second day of creation, which is the separation of good from evil).

Interpreted literally, the entire Samson affair is nothing more than wild fantasy. Considered logically – and *symbolically* – it is of lasting worth.

The Book of Esther

Written at a time when the prophetic authors were only too aware of the tendency of rival nations to persecute those of Jewish faith, with their rather "revolutionary" view of but one invisible God, it is not surprising to find that the Book of Esther opens with a story that stresses the value of an open mind – a virtue which, it was hoped, would become much more universal. One other basic theme focuses upon the moral that, as threatening as persecution of the Jewish race may sometimes be, as long as they are really in search of the Truth, God will protect them and assist them to overcome their enemies. These two primary morals may be described as follows:

The Virtues of an Open Mind

The *"one hundred and twenty-seven* provinces," over which King Ahasuerus is reported to have reigned (Esther 1:1), are a symbol of good government – inasmuch as his rule permitted one to recognize but one God, separate good from evil, and be free to live the sevenfold sense of the days of creation. That in "the *third* year of his reign" the king displayed the "riches of his royal glory" for a period of 180 days (Esther 1:3-4) is most instructive. Translated, we have the message that the king was mature spiritually and that, in the course of recognizing one true God, he strove to seek the Truth and separate good from evil. (At times the "four" is used in combination with the sense of the "second" day of creation to give the "eight" aspect.) The *seven* chamberlains who conveyed a command to Queen Vashti may be taken to represent the voice of conscience. The subsequent refusal of the queen to comply with the king's request symbolizes that state of mind which "knows it all" and will not bother to listen. Adherence to such an attitude is bound to lead to one's downfall – a point which is demonstrated when the queen is replaced by one blessed with the virtue of open-mindedness: Esther.

The symbolic role portrayed by Esther is rather evident to one versed in the system of biblical symbolism. Not only was she the most highly regarded of virgins (throughout the Scriptures a "virgin" is used as a symbol of virgin or uncontaminated thought), but the seven maidens whom she acquired are symbolic of the sevenfold process of spiritual advancement – which only an open mind is able to accept. The purifying treatment which she underwent for twelve months, before being presented to the king, conveys the message that one who is blessed with such virtue will indeed possess a philosophy endowed with authority. Moreover, the fact that Esther was taken to the king in the tenth month symbolizes that she was following the dictates of conscience; and the knowledge that it was in the seventh year of the king's reign when this event took place imparts the information that Esther's spiritual training was now complete and her destiny about to be fulfilled: she was made queen.

Those who Seek the Truth will not Perish

After stressing the importance of possessing an open mind, the narrative continues with a moral aimed at illustrating the serious consequences which may arise if immature minds elect to persecute individuals simply because they hold different religious beliefs.

The plot itself is built around a man named Mordecai (representing the Jewish faith) who was disliked by an official by the name of Haman (depicting a sense of ignorance). The *ten thousand* talents of silver which Haman offered in return for the king's edict "to destroy, to slay, and to annihilate all Jews, young and old, women and children, in *one* day, the *thirteenth* day of the *twelfth* month" (Esther 3:13) impart the meaning that the aspirations of a "Haman" are quite in contrast to the dictates of a healthy and enlightened conscience. That the Jewish inhabitants should be destroyed at such specific times may be taken to symbolize that this action would fail to recognize the one true God ruling the universe, would be the opposite of insight into the Truth put into practice, and be quite in contrast to a truly divine Government.

When Mordecai learned what was contemplated, he was naturally very much concerned and humbled himself before God (Esther 4:1), to the extent that he recognized his own deficiencies and sought the help of one higher than himself (the action of rending one's clothes and putting on sackcloth and ashes is used frequently in the Scriptures as an indication of humility). In a similar manner it is reported that at Esther's bidding all Jews underwent a *three-day* fast – symbolizing a divorcing of one's thoughts from matters of a material nature, and thereby allowing a more realistic perspective to manifest itself.

After a measure of insight into the Truth had been obtained (i.e., on the *third* day Esther put on her royal apparel), a banquet was then arranged at which the question of Jewish persecution would be settled one way or the other. On the *second* day at the banquet (Esther 7:2), good was finally separated from evil, inasmuch as Haman was himself hanged upon the very gallows which he had just built to hang Mordecai. That the gallows are described as being *fifty* cubits high symbolizes the opposite sense of the "fifth" day of creation, which is abundance of life. Invariably, one with evil tendencies must be expected to perish at the hands of materialism.

The *ten* sons of Haman who were also hung (Esther 9:14) convey the moral that those who would follow the policies of a "Haman" would just as surely perish. In addition, the various recorded "slayings" by the Jews (Esther, chapter 9) symbolize that those with spiritually immature minds cannot hope to prevail, but must fall before an earnest search for Truth. Thus the *three hundred* men slain on the *fourteenth* day of the month (Esther 9:12) will be seen to symbolize that a false perspective of matters must succumb to those who would seek the Truth.

The Trials of Job

The mystery as to why some righteous individuals may be observed to experience greater suffering and misfortune than others of a less righteous nature was puzzling to ancient minds (as it still is with many today), who found this difficult to reconcile with belief in a Just God. One explanation offered for this enigma was the notion that any illness or misfortune is really a direct result of our own sins. (This interpretation is, of course, quite untenable, since to believe that a specific misfortune is punishment for a specific sin is to deny the vital principle of *free will* – which alone is responsible for the introduction of chaos and imperfection!) However, the writers of this story appear to have been aware of the situation, repudiating such a notion with the narrative of Job's trials and tribulations – the moral of which is that it is a great mistake to attempt to link misfortune to one's level of righteousness. (It is interesting to note that many priest/authors of the Scriptures sought to blame much misfortune on a refusal to accept their views.)

The favorable symbolic role portrayed by Job is obvious. Not only is he described openly as a perfect and upright man, but the symbolic numerals characterizing his sons and daughters and certain worldly possessions further stress this point. For instance, the *seven* sons and *seven thousand* sheep together symbolize that Job possessed a true sense of the seven days of creation. His *three* daughters and *three thousand* camels depict a measure of insight into the Truth and of the desire to put such to good use.

Although he was subjected to two separate trials by Satan (meaning *ignorance*), the moral conveyed is essentially the same. The first trial concerns *four* false stories involving destruction of all that Job cherished (Job 1:13-19). The fact that Job dismissed these four false charges symbolizes that he was earnest in his search for Truth, since he did not succumb to a false line of thought which would accuse God of an injustice simply because it might appear so in the imperfect eyes of mortal man.

The second trial to which Job was exposed so undeservingly was a degree of physical suffering (Job 2:7). The *three* friends who would philosophize that his misfortunes must be directly related to his level of righteousness symbolize those well-meaning but misinformed individuals who would uphold such a false view. That Job's three friends "sat with him on the ground *seven* days and *seven* nights" (Job 2:13) before offering their opinions may be taken to imply a state of imperfection (i.e., the "seven" used in its opposite sense), and thus could not be allowed to influence one more spiritually enlightened – namely, Job. Passing through periods of intense remorse, in the end Job's faith prevailed and he wisely refused to consider God as being anything but righteous, even though he had been exposed to tribulations of which he was not deserving. The sacrifice of "*seven* bulls and *seven* rams" (Job 42:8), which Job's three friends were required to offer as a result of their folly, conveys the message that they were to exert every effort to perfect themselves.

The list of Job's assets, as described at the conclusion of this narrative, is rich in symbolism. Also of interest is the statement that "after this Job lived a *hundred and forty* years, and saw his sons, and his sons' sons, *four* generations" (Job 42:16), as it conveys the message that all who would seek the Truth would be successful.

Some Examples from Daniel

As an example of the work of later authors the episodes of Daniel contain many dramatic allegories depicting the irresistible power and dependability of Truth. A few instances of this rich symbolism are described as follows:

The Stone which Smote the Image

In the second chapter of Daniel we have the story in which King Nebuchadnezzar's dream is interpreted by Daniel. This dream concerns a great image which was broken into pieces by a *stone* that was "cut out by no human hand," and symbolizes the destruction of a false and mortal state of mind by the *appearance of Truth*. This great image is described as a composition of *five* distinct material substances or mixtures (gold, silver, brass, iron and iron/clay), and may be taken to symbolize a misleading philosophy based solely upon the five physical senses – exactly opposite to the abundant life of the fifth day of creation. Just as the image epitomizes materialism, so the *stone* which is able to destroy it symbolizes that precious insight that comes to us as *conscience* – revelation which is not defiled by the tradition of imperfect man. It is in truth "cut out by no human hand." If only humanity would recognize the need to follow one's own conscience, instead of blindly accepting the views of others, then perhaps one day it might well be said that a "stone" has become a "great mountain" and has "filled the whole earth."

Shadrach, Meshach and Abed-nego

We next come across another significant allegory in which *three* Hebrews (Shadrach, Meshach and Abed-nego) are cast into a furnace which was heated *"seven* times more than it was wont to be heated" (Daniel 3:19). This crisis came about as the result of their refusal to worship King Nebuchadnezzar's golden image, "whose height was *sixty* cubits and its breadth *six* cubits" (Daniel 3:1). In essence, they refused to worship the false and materialistic mode of thought so expressed by the sixfold aspect of the golden image. After thrusting these three men into the seven-times heated furnace it is recorded that "I see *four* men loose, walking in the midst of the fire, and they are not hurt; and the appearance of the fourth is like a son of the gods" (Daniel 3:25). Put into simple language, we have the moral that those who possess an insight into spiritual matters (i.e., the three Hebrews) need not fear the ravaging effects of materialism (i.e., the furnace that was seven times hotter – the opposite or "counterfeit" sense of the seven days of creation), since one obeying conscience (symbolized by the appearance of a *fourth* man – a sense of seeking the Truth) can never die spiritually and may thus be called a "son of the gods."

Daniel in the Lion's Den

The well-known episode of Daniel being cast into a lion's den, from whence he is reported to have emerged untouched, is obviously another allegory. The moral in this case is much the same as that of the furnace incident. Truth alone is able to provide real safety and deliverance from the adverse influence of materialism. The *five* classes of authorities which were instrumental in getting King Darius to sign a decree stating that "whoever makes petition to any god or man for *thirty* days, except to you, O king, shall be cast into the den of lions" (Daniel 6:7), represent that false state of mind the five physical senses are only too apt to foster. The decree itself, which dictated that no one was to ask a petition for *thirty* days (an obvious "counterfeit" of the spiritual meanings of the numerals "three" and "ten"), symbolizes that one is required to worship human idol (i.e., the king). Daniel's refusal to obey such a decree was to be expected, since he possessed insight into the Truth by making "his petition *three* times a day" (Daniel 6:13), and anyone with such vision could never obey the laws of spiritually immature authorities. Daniel's "miraculous" deliverance from the "lions" (a symbol of mortal actions) illustrates the virtue of possessing a philosophy based upon spiritual contemplation; whereas the subsequent perishing of his accusers at the hands of the very same "lions" signifies that a philosophy having a materialistic basis has no real life or substance and will therefore perish of its own accord.

The Story of Jonah

The impressionable story of Jonah and the whale is clearly another instance of the prophetic writers resorting to fiction in order to express a moral or principle. Essentially, the message imparted by this narrative stresses the obligation to follow conscience, even if it should be less than enlightened.

The city of Nineveh represents those individuals who, while they would readily accept the Truth if they were aware of it, nevertheless live in a state of ignorance. (In Jonah 4:11, it is reported that they could not discern "their right hand from their left.") Jonah, aware of this unfavorable prospect, disobeyed the command to preach to the people of Nineveh and instead attempted to get away from conscience: he "rose to flee to Tarshish" (Jonah 1:3). As is always the case with one of a really conscientious nature, it is impossible to find peace of mind if the dictates of conscience remain unsatisfied. Hence it is reported that "the Lord hurled a great wind upon the sea" (Jonah 1:4). The final result of this crisis was that, in order to find "dry land," increased insight into the Truth was essential: "Jonah was in the belly of the fish *three* days and *three* nights" (Jonah 1:17).

With renewed vision, Jonah then faced the duty of preaching in Nineveh, which was reported to be a *three-day* journey. (The task before him was indeed one which required a high perspective in order to succeed.) Jonah's statement that "yet *forty* days, and Nineveh shall be overthrown" (Jonah 3:4) conveys the warning that unless the inhabitants were sincere in their endeavor to find and live the Truth, they would be made to suffer the consequences. The fact that the people of Nineveh were forgiven when they repented of their misdeeds, even though many were still incapable of distinguishing between right and wrong, imparts the moral that sin must really be defined as a personal disobedience of conscience – irrespective of caliber.

Summary

While we have presented a brief insight into some of the more popular characters and events of the Old Testament, of necessity many others of interest have been omitted.

For instance, in I Samuel, chapter 17 there is the colorful and inspiring story of David and Goliath, in which the latter is depicted as a symbol of materialism. (It is stated that the height of Goliath was *"six* cubits and a span;" that the weight of his coat of mail was *"five thousand* shekels of brass," and that his "spear's head weighed *six hundred* shekels of iron.") That Goliath (i.e., materialism) was slain or overcome by a *stone* wielded by David is not surprising, since a "stone" is often used as a symbol of the *appearance of Truth,* which will slay any sense of evil.

Then again in II Kings, chapter 1 there is the episode of Elijah commanding fire to come down from heaven. The symbolic nature of this allegory should be obvious even to one not versed in the system of biblical symbolism; for to believe that God would intervene in such dramatic fashion to destroy *one hundred* men who could not be deemed any worse than the average, is totally unthinkable! Actually, the instructive moral behind this story, which centers around a reported persecution of Elijah because he dared to criticize King Ahaziah for his lack of monotheistic faith, is simply that those who fail to recognize one God will surely suffer as a consequence. For example, the first two attempts to take Elijah resulted in the "death" of two groups of *fifty* men (symbolizing mortal thought); it was only after a *third* attempt (when insight into the Truth was added) that their lives were secure.

In spite of a somewhat restricted analysis of the Old Testament, it will nevertheless be seen how futile – and foolish – it would be to attempt to reduce the various passages to mere physical events. (Some characters and incidents are obviously entirely symbolic; with others it is difficult to tell if some literal consideration might not be justified.) However, once it is realized that it is only *Principle* which is of fundamental and lasting importance, it follows that the question of whether a given narrative is fact or pure fiction is, of itself, really of minor consideration. What is important is that sentiment must not prevail over the revelations of science and logic.

8

The New Testament

The Enigma of Satan – The Jesus/Christ Paradox – The Genealogy
and Birth of Jesus – Preparation for His Mission – Spiritual
Significance of the "Miracles" – The Crucifixion
and Resurrection – The "End of the World"
Discourse – The Book of Revelation –
Words of Lasting Wisdom

Although of a generally higher nature than many documents of the Old Testament, and while dealing with the life and teachings of a most dynamic personage, it remains that the New Testament is primarily a *continuation of earlier symbolic and spiritually significant writings!* To not a few earnest seekers of Truth, it is bound to come as somewhat of a shock to learn that traditional interpretations are in need of considerable revision; but such is certainly the case. Indeed, even the basic titles of "Satan" and "Christ" demand an explanation that is far different from what has been offered by orthodox theology.

The Enigma of Satan

Throughout the ages the identity of Satan has proven to be a source of great embarrassment to many theologians. For while they have been inclined to look upon Satan as a sort of supernatural being and opponent of God, they have been quite unable to explain a rather fundamental mystery concerning his very existence. In fact, such theology is guilty of holding two highly conflicting points of view; for if God is really omnipotent, *then why does He simply not destroy Satan?* Obviously, any doctrine which would consider Satan to be a spiritual entity of a supernatural nature is quite ridiculous; and the very existence of such a belief may once more be traced to a distorted biblical interpretation.

Actually, the devil is nothing more than a particular *symbol* which was introduced into the Scriptures for the purpose of illustrating a *lack of Truth!* The entire problem of evil is but a false sense of reality. Indeed, is the symbolism portraying lack of the Truth not perfectly clear when we recall Jesus' reference to the devil? "He was a murderer from the beginning, and has nothing to do with the truth, because there is no truth in him. When he lies, he speaks according to his own nature, for he is a liar and the father of lies." (John 8:44).

What we might like to consider so conveniently as being the works of the devil should really be attributed to the results of our own selfish desires – our own state of imperfection! Whether or not it is realized, all sin originates from an unbalanced love of oneself; *for selfishness is at the root of all evil!* As we grow older, it is true, it may manifest itself in many different ways, but nevertheless all of the evil thoughts which may materialize in our minds can be traced to this one source. This being so, then we must certainly admit that our self-centered minds possess all the cunning and deceitfulness that the Scriptures would appear to ascribe to Satan. For it is none other than our own ignorant and selfish state which is ever working, often with great deception, to mislead us and blind us to the Truth.

If we would attempt to lead a sinless life, we must be made to understand that Satan is simply *our own inherently selfish nature;* for once we do, it will enable us to meet temptation on more equal terms. Whenever temptation arises (no matter in what form), we should first make an honest attempt to ascertain if our own selfish interests might not play a far greater role than we would like to believe. If only we would do this, then the chances of deceiving ourselves would be much reduced. We will have succeeded in bringing temptation out into the open, where it can then be seen in its true perspective.

The Jesus/Christ Paradox

Our fundamental problem is one which orthodox Christianity as a whole claims to understand; but when investigated, its extremely dogmatic explanation is seen to end up in a rather obscure and even contradictory situation. In short, who (or what) is a "Christ?" What is the true status of "Jesus," that historical personage whom so many hold in such high esteem?

On the basis of orthodox views, we are apparently to believe that Jesus was a one and only "Son of God" – that no other comparable entities ever existed. This notion can only be described as highly naive, for even if we were to accept the traditional idea that Jesus constituted our world's one and only "Saviour," it sadly fails to take into consideration all the countless numbers of *other worlds* similar to our own which have existed in the past, exist now and which are bound to exist in the future! Surely we cannot be so vain as to deny these other worlds the light of a "Christ" such as ours – or do we have so much as the slightest idea as to the true nature of the spiritual cosmos? Clearly, this popular concept of a "Saviour" is not a very realistic one and breaks down completely under the slightest reasoning.

Upon examination of the Bible itself, we encounter a most confusing and contradictory state of affairs. On the one hand, we have the statement "I and the Father are one" (John 10:30), with the rather curious doctrine of a so-called Holy Trinity based upon the belief that Jesus was indeed *equal* to God; while on the other hand we have numerous passages which impress upon us the understanding that God is *greater* than Jesus: "the Father is greater than I" (John 14:28); "not my will, but thine, be done" (Luke 22:42); "I do nothing on my own authority but speak thus as the Father taught me" (John 8:28); "If you keep my commandments, you will abide in my love, just as I have kept my Father's commandments and abide in his love" (John 15:10); "I can do nothing on my own authority; as I hear, I judge; and my judgment is just, because I seek not my own will but the will of him who sent me" (John 5:30); "And Jesus said to him, 'Why do you call me good? No one is good but God alone.'" (Luke 18:19).

The solution to this contradiction (which, incidentally, orthodox theology has yet to explain) is simple in essence. Recalling that the Scriptures deal primarily with principles, and that historical events are of secondary importance, it should not come as a surprise to learn that Jesus was described initially – *and symbolically* – in terms of a "dual" personality. That is to say, when referring to actual human status of Jesus it must certainly be acknowledged that "the Father is greater than I" (John 14:28), since we are dealing with an entity characteristic of mankind – a "son of man." In contrast to this imperfect state of mind we have the "Christ" or "Son of God" state, which is really a *title or goal* in that it *symbolizes* the attainment of a state of Perfection which is none other than God Himself:

"I and the Father are one." (John 10:30). Thus it will be seen that, throughout the course of his earthly ministry, Jesus was striving to demonstrate and live the *hypothetical* goal or ultimate "Christ" state of Perfection. This principle of attempting to reach such a priceless goal we may term the **"Christ-idea"** – for essentially, "Christ" is a *spiritual idea or ambition* of which it may well be said that "I am the way, and the truth, and the life; no one comes to the Father, but by me." (John 14:6).

It will, therefore, be concluded that Jesus himself was a very human being, and all attempts to picture him as anything else will be seen to be rooted in sheer fantasy.

The Genealogy and Birth of Jesus

As one might suspect, the attempts of Matthew and Luke to trace the genealogy of Jesus back through the ages are not to receive a literal interpretation. On the contrary, they serve to convey two messages of a distinctly spiritual nature. Is it not a rather strange coincidence to note that in the genealogy listed by Luke, *Jesus happens to be the 77th name?* Recalling the symbolic meaning of the numeral "seven," the reason for this is obvious: it symbolizes that Jesus was "fulfillment" – that he was indeed capable of making an impressive demonstration of the *Christ-idea*. In the genealogy that is recorded in Matthew, is it not equally significant that there are *three* very distinct groups, each of which contains *fourteen* ("ten" plus "four") generations (Matthew 1:17)? In other words, Matthew is conveying the message that, in the course of his endeavor to seek the Truth (i.e., the "four"), and apply such light to human experience (i.e., the "ten"), Jesus would inevitably give a famished world welcome insight into the nature of the spiritual cosmos (i.e., the "three" aspect).

Incidentally, it must be clear that efforts to trace the genealogy of anyone back thousands of years at that early date can only be regarded as absurd; for even today it is practically impossible to track our own ancestry back more than a few hundred years. So we see that all things considered, there is really no reason to suppose that the authors of such genealogies actually believed in the historical truth of their stories, although it would no doubt suit their purpose if the masses could be persuaded to give them credence.

The recorded birth of the babe Jesus contains many examples of symbolism, of which the following instances will be found to convey instructive morals:

The New Testament

"Behold, a virgin shall conceive and bear a son." (Matthew 1:2)

The real meaning behind this "virgin birth" story involves recognition that the "Christ-idea" (i.e., the principle of rising to a state of Perfection) will only be born to one with "virgin thought" – to one who has not been contaminated with false theology.

"And she gave birth to her first-born son and wrapped him in swaddling clothes, and laid him in a manger; because there was no place for them in the inn." (Luke 2:7)

Here we have a beautiful example of symbolism. The birth in a "manger" symbolizes that there is never any room (or time) for the "Christ-idea" to enter into the hearts of well-to-do people concerned with worldly affairs (i.e., the "inn" state of mind); for it can only be born to those of a humble or lowly "manger" state of mind.

"Now when Jesus was born in Bethlehem of Judea in the days of Herod the king, behold, wise men from the East came to Jerusalem, saying, 'Where is he who has been born king of the Jews? For we have seen his star in the East, and have come to worship him.'" (Matthew 2:1-2)

This "star in the East" represents the guiding light of conscience, which any wise man will follow.

"Then Herod summoned the wise men secretly and ascertained from them what time the star appeared." (Matthew 2:7)

Herod may be seen to depict that mortal state of mind which always wants to take things of a higher spiritual nature and reduce them to material terms concerned with such trivialities as time, persons and places.

"Then, opening their treasures, they offered him gifts, gold and frankincense and myrrh." (Matthew 2:11)

This story of the wise men presenting three gifts to the infant Jesus may be interpreted to mean that he would indeed be given insight into the Truth.

"Now when they had departed, behold, an angel of the Lord appeared to Joseph in a dream and said, 'Rise and take the child and his mother, and flee to Egypt.'" (Matthew 12:13)

The fact that it was imperative to flee from Herod for a time symbolizes that, in the face of evil, it is sometimes advisable to take the newborn spiritual idea and hide its light until developed sufficiently to assure complete triumph.

"Then Herod ... sent and killed all the male children in Bethlehem and in all the region who were two years old or under." (Matthew 2:16)

This "slaying" of young children is similar to that reported at the time of the birth of Moses; it symbolizes that the world is forever seeking to destroy those who would want to separate good from evil.

Preparation for His Mission

From the recording of the birth of Jesus until the actual start of his mission, we encounter some interesting examples of symbolism. First of all, it will be noted that the only mentioned incident of the childhood of Jesus is at the age of *twelve,* where he was lost for some *three* days and eventually found "in the temple, sitting among the teachers, listening to them and asking them questions" (Luke 2:46). Recalling the symbolic meaning of the numeral "twelve," the actual circumstances under which he was found the "third" day, and Jesus' explanation that "I must be in my Father's house," we arrive at the following interpretation: In order to demonstrate the universal scope and authority of divine Principles to others, we must also be "three" days in our Father's house (i.e., we must first obtain insight into the Truth).

After this episode at the age of twelve years we next hear about Jesus at the age of *thirty* years ("three" times "ten"), when he is reportedly baptized preparatory to setting out upon his preaching mission. Whether Jesus really started his preaching at the age of thirty years (Luke 3:23) is quite unimportant; the message intended is that Jesus possessed insight into the nature of spiritual matters (i.e., the "three"), and was subsequently about to apply this insight to human experience (i.e., the "ten").

The very act of baptism is one which is obviously misunderstood by theologian and layman alike. Introduced into biblical literature as a *symbol* of imparting to one some measure of insight into the Truth, the water merely symbolizes receipt of a superior philosophy. It is most unfortunate that this symbol has since become converted into an empty ritual. We now witness priests going through the motion of sprinkling actual water upon individuals in the vain belief that their efforts are able to raise such souls in the esteem of God. As a matter of fact, not only are adults subjected to a pointless physical ritual, but the general intention seems to be to baptize tiny infants quite incapable of thought! Clearly, a mere symbol has become distorted beyond all reason. Only in the sense that a person has received a more realistic insight into the Truth may it be said that one has been truly baptized; all else is of no avail.

After his baptism, signifying that he indeed possessed insight into the Truth, Jesus then experienced *three* temptations by the devil. Here we have an instance in which the symbol "three" is used in its opposite sense; for it signifies that our own imperfect nature (i.e., the "devil") is forever tempting us with a false perspective. From the report that Jesus "fasted *forty* days and *forty* nights" in the wilderness (Matthew 4:2), we may conclude that he was earnest in his desire to seek and live the Truth. (The term "wilderness" is used throughout the Scriptures as a place where the material sense of things gives way to the spiritual.) The information that "afterward he was hungry" really means that Jesus was hungry or desirous of going about his Father's business.

Coming to the actual mission itself, it is most interesting to note that Jesus chose *twelve* apostles – a number symbolizing the authority and universal extent of divine Principles. An investigation of the calling of the apostles also reveals much of a symbolic nature. For instance, in Matthew (4:18-22), is it not instructive to read that Jesus called *four* followers initially (Peter, Andrew, James and John); while in Luke (5:1-11) there are also *four* mentioned on a "fishing" expedition (Jesus, Peter, James and John)? Can we not discern here the same symbolic message as in the saga of Abraham where *four* people are listed as setting out for the "land of Canaan" (i.e., in search of the Truth)? What we need today is more individuals who would make it a priority to become "fishers of men."

Spiritual Significance of the "Miracles"

Invariably, the so-called "miracles," which center around the life of Jesus, are just as thoroughly embedded in allegorical structure as the many earlier instances of the Old Testament. Upon applying our knowledge of biblical symbolism to certain of these purported events we may derive the following interpretations:

The Feeding of the Multitudes:

As misguided theologians of the past first saw fit to teach, so many of our present-day churches exhibit little improvement and still tend to look upon Jesus' "feeding of the multitudes" in a mere earthly and materialistic sense. Taken literally, such an incident is not only incomprehensible, but is quite contradictory with respect to an earlier occasion where Jesus refused to turn a stone into bread that physical wants might be satisfied! On the contrary, far from feeding the multitudes with physical food, Jesus fed them with *spiritual food!*

Can we not see that those instances of dispensing "bread" refer not to relatively unimportant physical food, but to the *Word or Bread of Life?* "Jesus said to them, 'I am the bread of life; he who comes to me shall not hunger, and he who believes in me shall never thirst.'" (John 6:35). "This is the bread which came down from heaven, not such as the fathers ate and died; he who eats this bread will live for ever." (John 6:58). Is not the symbolic aspect of the "bread" quite unmistakable when we further read Jesus' warning not to eat of the leaven of the Pharisees and of the Saducees? "'How is it that you fail to perceive that I did not speak about bread? Beware of the leaven of the Pharisees and Saducees.' Then they understood that he did not tell them to beware of the leaven of bread, but of the *teaching* of the Pharisees and Saducees." (Matthew 16:11-12).

Upon looking deeper into the two occasions where Jesus is reported to have fed the multitudes, the whole affair becomes crystal clear once we apply our knowledge of the symbolic numerals. For instance, in Matthew, chapter 14, it is recorded that *five* loaves and *two* fishes were divided among *five thousand,* from which *twelve* baskets of food were later gathered. As the "five" is used here in the sense of the fifth day of creation (i.e., fullness of Life) and the "two" in the sense of the second day (i.e., a separation of good from evil), it will be seen that when such a philosophy has been digested by those desirous of obtaining a higher and more abundant sense of Life (i.e., the "five thousand"), it will result in the Word being turned into a successful demonstration of the power and universal extent of divine Principles (i.e., the "twelve" baskets of fragments).

In Matthew, chapter 15, we have a slightly different rendition. In this case it is said that *seven* loaves and a few fishes were distributed among *four thousand* men who had fasted *three* days after ascending high ground (symbolizing exalted thought) and that *seven* baskets of food were left over. Reduced to simple language, we have the message that Jesus imparted to those who were really desirous of seeking the Truth (the "four thousand" who had fasted "three" days) an abundant sense of Life (the "fishes") and insight into the spiritual meaning behind the seven days of creation (the "seven" loaves) which, if given time and utilized properly, are bound to lead to multiplication, fulfillment and a state of Perfection (the "seven" baskets).

The Water Turned to Wine:

This traditional "miracle" (John 2:1-10) is obviously another allegory used to illustrate a principle of lasting value. The overall moral to be drawn from this story may be stated as follows: Whenever the Word of Life (symbolized by the water) enters into a false sense of reality (epitomized by the *six* waterpots), it will effect a change and produce inspiration terminating in the recognition that Love is supreme (*wine* is used throughout the Scriptures as a symbol of the inspiration of Love). In other words, to the extent that we let the *Christ-idea* enter our hearts it will inspire us and lift our imperfect concepts to a truer and more filling sense of Love.

The marriage feast itself, of course, does not refer to the usual union of two members of the opposite sex, but rather to the "wedding" of human minds to a higher and more advanced outlook upon life. Hence it is reported that on "the *third* day there was a marriage at Cana in Galilee" (John 2:1); the participants indeed received an insight into the nature of spiritual matters.

The Man Healed of the "Withered Hand:"

This particular episode, in which a man's withered hand is alleged to have been restored physically, must likewise submit to reason. Does it not strike one as rather odd that the Pharisees, before whom this act was performed, failed to be impressed by such an incredible event? Would one not expect even the blackest sinner to immediately bow down in sheer fright and worship the performer of so vivid a miracle? Of course one would! But what was their reaction? They actually "went out and took council against him, how to destroy him" (Matthew 12:14).

In essence, the moral which the author of this story is attempting to convey is that the false and enslaving theology of the Pharisees, which would reduce an individual's duty to seek the Truth to a slothful reliance upon organization and blind obedience to the will of so-called authorities, had withered and deadened the people's sense of reality. Moreover, given the chance, the light of the *Christ-idea* (the progressive revelations of conscience) would eventually restore them to spiritual health. Thus we can readily see why the Pharisees were not impressed by any "physical" miracle but sought only to destroy Jesus, since his teachings were succeeding in opening the public's eyes to the necessity of *thinking for themselves* and to the utter futility of relying upon their withered philosophy. If only the world could free itself from the pernicious grip of some of its modern Pharisees!

The Raising of Lazarus:

Like other biblical accounts involving the restoration of sight and physical health, the raising of Lazarus is an allegory which illustrates a definite moral. To mortal eyes a man appears dead if he is inert physically; while in reality the only true death is one of a spiritual nature. Hence, although mortal man might consider one to be dead in the physical sense, a higher and more advanced outlook will disclose that the spiritual (i.e., one's level of righteousness) is never dead and is the only true reality and the only intrinsic source of Life.

Examining the story of the raising of Lazarus, we promptly come across a most interesting discovery; for he is reported to have "been in the tomb *four* days" (John 11:17). Clearly, the moral behind this particular example of symbolism is that, regardless of mere physical occurrences, if one has been earnest in his search for Truth he cannot be dead, since the spiritual is the only true source of Life and the only real death. Consequently, it will be seen that the *stone* which was taken away (John 11:41) refers not to any material stone, but to the removal of that false perspective which would consider physical death to be all-important (the "stone" is used here in its opposite sense). How infinitely more serious is it to be spiritually dead, than to undergo but a physical and temporary death!

Can we not see that such instances, which orthodox theology would take in an earthly and physical sense, really refer to the much more important spiritual sense? Is it not clear that throughout the course of his mission Jesus instilled into people a more realistic sense of Life and brought many who were sick spiritually (or even "dead" to the Truth) to real Life and spiritual health? No wonder it is reported that as many as touched the hem of his garment "were made well" (Matthew 14:36), since whenever one touches or comes to accept the *Christ-idea,* any spiritual sickness (or "blindness") will instantly become past history.

The Walking on the Sea:

Among the traditional "miracles" are those of Jesus walking on the sea and the stilling of the tempest. However, just like so many other distorted theological concepts which erroneously neglect the spiritual in favor of the physical, a drastic revision is again indicated. For instance, the "sea" incident is not to be taken as referring to any physical body of water at all; but rather the term "sea" is used over and over again in the Scriptures to symbolize exposure to *adverse and worldly influences.* Accordingly, it will be seen why Jesus was so successful in stilling the "tempest," since error and a materialistic mode of thought will naturally vanish before the virtues of a superior philosophy.

In connection with the episode of his walking on the sea, is it not instructive to note that Jesus came walking to his disciples "in the *fourth* watch of the night" (Matthew 14:25)? Actually, the reason for the disciples finding themselves "beaten by the waves; for the wind was against them" (Matthew 14:24) was that they had not gone with Jesus up into the mountain to pray. That is to say, they had not exerted the necessary thought to grasp or digest the full meaning of the "spiritual feast" which had just taken place. Nevertheless, in spite of periods of temporary regression, if in our hearts we really desire to draw closer to the Truth we will not be denied (i.e., Jesus, representing the *Christ-idea,* is reported to have drawn toward the disciples in the *fourth* watch). The subsequent incident, in which Peter is said to have attempted – unsuccessfully – to duplicate Jesus' feat of walking on the sea, conveys the moral that one must have faith in the *Christ-idea* and not succumb to the ravaging effects of an imperfect and hostile world.

The Restoration of the Severed Ear:

The incident in the garden of Gethsemane, in which the slave of the high priest is first supposed to have his right ear cut off by one of the apostles and then to have it so miraculously restored by Jesus (Luke 22:50-51), is so obviously meant to be interpreted symbolically that it is puzzling why our theologians would ever think otherwise. Can anyone honestly believe for one moment that these soldiers could have witnessed so spectacular a miracle and then promptly turn around and arrest Jesus? Hardly likely! Much rather would they have quaked with fear of what they were about to do.

In reality, the moral behind this story may be stated as follows: Do not be a foolish zealot who, like Peter, would in one instance pass up a golden opportunity to gain insight into the nature of God ("Could you not watch *one* hour?") and then when a crisis arises would prefer to rely upon mere physical powers and "cut off somebody's ear." An understanding of Truth is a far greater asset to one than any physical superiority; in the face of temptation to resort to physical means in order to solve our problems we would do well to remember this very point: "Then Jesus said to him, 'Put your sword back into its place; for all who take the sword will perish by the sword.'" (Matthew 26:52).

The Blinding of Paul for Three Days:

The story of Paul being "blinded" for *three* days, while on the road to Damascus, has naturally been the subject of another literal interpretation by misguided theologians who apparently can see nothing wrong with a God who "specially" interferes with nature. The truth about this narrative, of course, is that it was written to symbolize the dawning of the light (i.e., the *Christ-idea*) to Saul, who was subsequently renamed Paul, just as certain other biblical characters were similarly renamed in order to signify their new relationship with God. The "light from heaven" which "flashed about him" (Acts 9:3) refers to no mere physical light, but to the *enlightening revelations of conscience* which indeed inquired of Saul: "Saul, Saul, why do you persecute me?"

After Paul was *three* days "without sight, and neither ate nor drank" (Acts 9:9), it is recorded that "immediately something like scales fell from his eyes and he regained his sight" (Acts 9:18). In other words, Paul had passed through a state of mental conflict only to emerge triumphant and in the possession of a much higher outlook. Whatever the eventual consequence, the conversion of Paul stands as a remarkable example of repentance and noble sacrifice for the sake of conscience.

The Crucifixion and Resurrection

Like other biblical narratives written primarily to illustrate a moral or principle, the events surrounding the crucifixion and alleged resurrection on the *third* day are no exception. For while they may indeed be based upon the actual death of Jesus, the use of symbolism is nevertheless quite

The Last Supper:

As proved to be the unfortunate case with regard to baptism, so theology has again seized upon symbolic writings and concocted another senseless ritual. First of all, the "large upper room furnished" which a man "carrying a jar of water" showed Peter and John (Luke 22:10-12) refers not to any mere physical event, but rather symbolizes that the "water" (i.e., the Word of Life) will lead one to a state of exalted thought – to a superior outlook as expressed by the "large upper room furnished." The last supper itself, at which Jesus is reported to have distributed "bread" and "wine" to his apostles, is not to be taken literally as some would have us believe. Instead, it was a distinct sense of Life and the inspiration of Love which Jesus imparted to his disciples. Hence the words "this do in remembrance of me" really means to "do likewise" and impart to others a higher sense of Life and Love. To believe that the popular physical ritual of a so-called Holy Communion is capable of imparting to one anything of value, as a result of the presence of the bread and wine, is quite ridiculous. It is strongly opposed to both common sense and to a Just Creator, and certainly has nothing at all to do with genuine religion.

Judas Betraying the Christ-idea:

Taken literally, the Judas episode would convey the horrifying impression that there is such a thing as predestination. (To believe that our very actions are predetermined is to deny the principle of free will!) Actually, Judas is portrayed as a symbol of imperfect *mortal thought,* which is ever bound to betray or renounce the Truth (i.e., the *Christ-idea*). Is it not instructive to note that Judas betrayed "Christ" for *thirty* pieces of silver, symbolizing exactly the opposite of putting spiritual insight into practice? How often are we ourselves guilty of betraying the *Christ-idea* for some mere transitory and material advantage!

The Three Denials by Peter:

As a sort of sequel to the moral of Judas betraying "Christ" for *thirty* pieces of silver, we have the *three* denials by Peter: "Jesus said to him,

'Truly, I say to you, this very night, before the cock crows, you will deny me *three* times.'" (Matthew 26:34). In this case the "three" is also used in its reverse sense, signifying a spiritual lapse or regression. The fact that *three* denials were to take place before morning symbolizes that before the light finally dawns upon one there will be occasions when the *Christ-idea* will be rejected and forgotten. Clearly, we have a great contrast between Peter's three denials and the moving drama in the garden of Gethsemane, where Jesus prayed *three* times that the "cup" might pass from him. (This "cup" represents the agony of seeing the spiritual ideas which he so highly cherished be misunderstood, ignored and defiled by mortal minds.)

The Releasing of Barabbas:

Barabbas is introduced into the Scriptures as a symbol of highly imperfect mortal thought. Hence the choosing of Barabbas over the "Christ" is not at all surprising, for even today in this present age the world still seems to prefer the "Barabbas" state of mind to an ideal which requires sacrifice and the expenditure of much effort. How frequently have we all, at one time or another, been guilty of choosing the Barabbas state in preference to the far more demanding *Christ-idea!*

The Crucifixion:

As a sample of the rich symbolism, which permeates the biblical record of the crucifixion, we may cite the following examples:

"And it was the third hour, when they crucified him." (Mark 15:25)

It was certainly an hour devoid of spiritual vision (the "three" is used in its opposite sense)!

"And with him they crucified two robbers, one on his right hand and one on his left." (Mark 15:27)

The two others crucified with Jesus (making *three* in all) may be taken to signify what the world is ever threatening to do to one with a superior outlook. In addition, the two robbers themselves signify that the world is frequently unable to discern right from wrong.

"When the soldiers had crucified Jesus they took his garments and made four parts, one for each soldier." (John 19:23)

Jesus' garments represent divine Principle, and the action of dividing his garments into *four* parts signifies that they split up and thereby failed to recognize the Truth (the "four" being used in its reverse sense to convey failure to seek the Truth).

"But his tunic was without seam, woven from top to bottom; so they said to one another, 'Let us not tear it, but cast lots for it to see whose it shall be.'" (John 19:23-24)

The fact that they "cast lots" for his tunic "without seam" imparts the moral that many do not have a genuine desire to work for a higher perspective and so are willing to "take a chance" on whether or not the undivided Truth will come to them of its own accord. (Invariably, many individuals find themselves with a false sense of values simply because of reluctance to expend effort on behalf of a superior philosophy, choosing instead to follow the easiest course and accept whatever views happened to be the most popular.)

"And he said to him, 'Truly, I say to you, today you will be with me in Paradise.'" (Luke 23:43)

This reply by Jesus to the repentant thief on the cross expresses the moral that God is of a forgiving nature and that what will determine a person's future status, quite regardless of past deeds, is the actual state of one's soul at the time of death.

"It was now about the sixth hour, and there was darkness over the whole land until the ninth hour, while the sun's light failed; and the curtain of the temple was torn in two." (Luke 23:44-45)

There is indeed darkness at any "sixth" hour ("six" is often used as a symbol of mortal thought) which will last until the "three" (i.e., spiritual insight) has been added, making the "ninth" hour. With possession of genuine knowledge the "curtain of the temple" (a symbol of false and blinding theology) will in truth be "torn in two"; it will be utterly destroyed.

The Rising on the Third Day:

Clearly, what is meant by this rising on the *third* day is that the righteous nature of Jesus must assure a *spiritual resurrection*. What invariably took place, of course, was an instantaneous resurrection (time does not exist as such between incarnations) in which the soul of Jesus was permitted the higher manifestation so richly deserved. (In this instance the "three" symbolizes insight into the principle of spiritual immortality.) That this rising on the "third" day was to be considered symbolic is borne out most conclusively by the statement: "For as Jonah was *three* days and *three* nights in the belly of the whale, so will the Son of man be *three* days and *three* nights in the heart of the earth" (Matthew 12:40), *since the episode of Jonah and the whale is so obviously an allegory!*

Upon looking over the various stories concerning discovery of the "resurrection", we note that emphasis is placed upon the "stone" which was "rolled away from the tomb." The tomb represents imperfect (or "dead") human hearts; the *stone* is this time used in the reverse sense to signify the manifestation of a false philosophy. Hence, on the *third* day the appearance of spiritual insight will have served to roll away a false perspective from human hearts. The result of this rolling away of untruth was the realization that "He is not here; for he has risen" (Matthew 28:6). For the "Christ" is not a mere physical body tangible to physical senses; but rather the *Christ-idea* depicts a *spiritual goal* – of the process of rising to a state of Perfection – and is therefore capable of being perceived only by those inclined spiritually. The "angels" in white and "shinning garments," which are reported to have informed certain of Jesus' followers that he had risen, are introduced as "messengers of God" and may be taken to depict the revelations of conscience.

Three Appearances of the "Christ:"

As one might suspect, the *three* appearances of the "Christ" are really attempts to express a moral or principle. Essentially, they are intended to signify that the apostles of Jesus would be successful in their quest for Truth (symbolized by the *three* appearances of the "Christ" unto them).

Taking the Gospel of John as an example, these three appearances may be resolved as follows:

1) *The appearance to Mary Magdalene:* It is interesting that the initial reappearance of "Christ" (i.e., of the *Christ-idea*) should be made at the tomb where, moments before, there was nothing to see. Mortal mind is bound to search for things of a physical nature. It is only after light has been added and Truth has been separated from falsehood (epitomized by the *two* "angels in white") that one is

able to grasp that "Christ" is really an *Idea* or *Principle,* which no man can destroy and which must ever be with us even though we might have no physical or visual proof. Physical senses can visualize no more than material phenomenon; while conscience is able to perceive things of a much higher spiritual nature.

2) *The appearances to the apostles in a closed room:* The two "closed door" appearances, once with Thomas absent and once with him present some *eight* days later – signifying a search for Truth and success in separating good from evil (i.e., the "eight" or "four" times "two" aspect) – conveys two morals. First of all, the "closed room" symbolizes the same open-minded state of thought as does the "virgin birth" moral; for once we close our doors to all adverse external influences we must leave our hearts open for the appearance of the *Christ-idea* (i.e., the revelations of conscience). Secondly, the "doubting Thomas" episode expresses the moral that it is wiser to believe in principles revealed by conscience than to accept only what is perceived by our imperfect and limited physical senses.

3) *The appearance to the apostles who were fishing:* The fact that the apostles could not discern the *Christ-idea* until they had first cast their nets on the "right side of the boat" (John 21:6), may be taken to mean that one's thinking will bear no fruit unless such virtue is utilized correctly. The actual catch itself, "a hundred and fifty-three" large fish, is interesting inasmuch as it contains the basic symbols "one" and "ten" along with the "three" and the sense of the "fifth" day of creation. Reduced to simple language, we have the explanation that an abundance of Life is bound to be the reward of those who would seek to expound the virtues of the one true God.

The "End of the World" Discourse

To many, one of the more perplexing aspects of the Scriptures is the apocalyptic literature, which is commonly believed to consist of prophecy and to refer to worldly events of a physical nature. This is an unrealistic and unfortunate interpretation, for the real purpose of this form of writing is to symbolize events of a decidedly *spiritual* nature. Historical events are often interwoven with symbolism in an attempt to add authenticity to the text. In particular, the composer of an apocalyptic work will frequently endeavor to select an event in history which has already come to pass and then weave it into the story so that it will appear to be a genuine prophecy concerning the future – thereby obtaining much valuable prestige. (An example of this would be Daniel, which gives one the impression of being written in the sixth century BC and contains a glimpse into future events; while in reality it was composed near the middle of the second century BC, when such "prophecies" were already ancient history! Another instance would be what many religious leaders still consider to be a prediction of the destruction of the temple in Jerusalem in the year 70 AD – a "prophecy" written long after the event had occurred.)

Although often looked upon as a prophecy by Jesus involving the physical end of the world and the coming of "Christ," such passages really refer to *spiritual fulfillment* (i.e., the universal acceptance of the *Christ-idea*). A brief discussion of this symbolism follows:

"Truly, I say to you, there will not be left here one stone upon another, that will not be thrown down." (Matthew 24:2)

This supposed reference to the destruction of the temple does not concern a mere physical structure, but rather refers to the destruction of *false beliefs!* "Do you not know that you are God's temple and that God's Spirit dwells in you?" (I Corinthians 3:16.) The "stones" here are mentioned in their opposite sense, signifying the destruction of false beliefs. (In addition, there is likely a secondary hope that this passage will come to be viewed as a prophecy foretelling destruction of the temple by the Romans.)

"Tell us, when will this be, and what will be the sign of your coming and of the close of the age?" (Matthew 24:3)

This passage really refers to world acceptance of the *Christ-idea,* and to the end or destruction of the materialistic mode of thought in each one of us.

"For many will come in my name, saying, 'I am the Christ, and they will lead many astray.'" (Matthew 24:5)

How true! The world has most certainly been plagued with the false teachings of so-called religious "authorities."

THE NEW TESTAMENT

"And you will hear of wars and rumors of wars; see that you are not alarmed; for this must take place, but the end is not yet." (Matthew 24:6)

Along with several of the following verses, this passage means that universal acceptance of the *Christ-idea* will not be easy to a world so enveloped in materialism.

"And this gospel of the kingdom will be preached throughout the whole world as a testimony to all nations; and then the end will come." (Matthew 24:14)

Universal acceptance of the *Christ-idea* is quite impossible until all have received insight into the Truth.

"So when you see the desolating sacrilege spoken of by the prophet Daniel, standing in the holy place (let the reader understand)." (Matthew 24:15)

This passage (see Daniel 12:11) refers to the taking away of the "daily sacrifice," which in turn was a symbol of the very principle behind the spiritual meaning of the seven days of creation.

"Then let those who are in Judea flee to the mountains." (Matthew 24:16)

In order to escape the consequences of failing to understand the significance of the seven days of creation, one must "flee to the mountains" – one must turn to exalted thought in order to rise above the situation.

"Pray that your flight may not be in winter or on a sabbath." (Matthew 24:20)

Do not attempt to reach the Truth through the coldness and emptiness of materialistic theories, nor through false and dogmatic theology.

"For as the lightning comes from the east and shines as far as the west, so will be the coming of the Son of man." (Matthew 24:27)

The light of the *Christ-idea* will one day be universal and it will appear to all. (This coming of the "Son of man" may be taken to symbolize the highest possible perfection of the human race. Hence, the "Sons of men" must ultimately evolve to become the "Sons of God.")

"Immediately after the tribulation of those days the sun will be darkened, and the moon will not give its light, and the stars will fall from heaven, and the powers of the heavens will be shaken." (Matthew 24:29)

The Sun, Moon and stars are used here in their opposite symbolic sense and convey the message that *false systems or principles* are bound to fail, and must finally give way to knowledge that will indeed shake the "powers of the heavens."

"And he will send out his angels with a loud trumpet call, and they will gather his elect from the four winds, from one end of heaven to the other." (Matthew 24:31)

Those who have sought the Truth (i.e., the "four") will be justly rewarded.

"So also, when you see all these things, you know that he is near, at the very gates. Truly, I say to you, this generation will not pass away till all these things take place." (Matthew 24:33-34)

One possessing insight into the Truth will indeed know when salvation has come to pass; in fact, such virtue will be responsible for its coming to pass.

"As were the days of Noah, so will be the coming of the Son of man." (Matthew 24:37)

As the one is symbolic (i.e., Noah and his ark), so is the other!

"Watch therefore, for you do not know on what day your Lord is coming." (Matthew 24:42)

The intended moral is that we do not know when death will overtake us, and that we should strive always after righteousness that we might not be caught unaware.

The Book of Revelation

To those not versed in the system of biblical symbolism, the Book of Revelation must pose an enigma. As a matter of fact, so saturated with symbolism is it that it was very nearly never incorporated into church canon at all, since in the eyes of many early leaders – lacking the necessary key – it

The New Testament

naturally appeared quite obscure. Regrettably, while most of our present-day theologians have been forced to admit that this work is largely allegorical in nature, they too have failed to offer a realistic interpretation.

Upon looking over the extensive symbolism so utilized, one finds that there is hardly a verse which does not contain one or more of the symbolic numerals! For this very reason it is extremely difficult to escape the conviction that its author has purposely overworked these "sacred" numbers in order to impress upon us the knowledge that the Scriptures were never intended to afford an account of mere historical events, but are instead concerned with the far more important realm of the spiritual.

In reality, throughout the entire course of this apocalyptic text there is an attempt to stress – *by means of symbolism* – the inevitable triumph of spiritual enlightenment over the darkness of mortal thought and mortal actions. This view is well illustrated in the very first chapter: "Behold, he is coming with the clouds, and every eye will see him, every one who pierced him; and all tribes of the earth will wail on account of him." (Revelation 1:7). The statement that "he is coming with the clouds" means that the *Christ-idea* has been expressed in the Scriptures through symbolism; and the words that "every eye will see him, every one who pierced him" conveys the message that the *Christ-idea* must one day be accepted universally, even by those whose thought is so darkened that they would "slay" the very idea. That "all tribes of the earth will wail on account of him" informs us that it will not be easy for the world to give up many cherished doctrines and a materialistic philosophy which would appear to afford immediate and tangible rewards for one's effort.

Inasmuch as this work is imbued with so many examples of symbolism, we must be content with but an introductory insight. Passing by such interesting instances as the diabolical beast in Chapter 13, whose "number is *six hundred and sixty-six"* (symbolic of imperfect mortal thought); the *"hundred and forty-four thousand* who had been redeemed from the earth" in Chapter 14 (symbolizing that those who sought to find and obey the One Infinite Cause will in truth be identified with God); and the ultimate downfall of that great city of Babylon (also in Chapter 14 and referring now to the fall of Rome – *a symbol of mortal beliefs* – just as the falling of the walls of Jericho was used to symbolize the same annihilation of evil), we may take the following examples for brief comment:

215

The Messages to the Seven Churches:

The *seven* messages to the *seven* churches (Revelation, chapters 2 and 3) constitute an interesting specimen of biblical symbolism. Each message will be seen to correspond with the spiritual meaning imparted by one of the seven days of creation. This relationship may be stated as follows:

Message to Ephesus:

> This church is accused of neglecting its "first love" (i.e., the first light of conscience, symbolized by the *first* day of creation).

Message to Smyrna:

> We have here an illustration of the sense of the *second* day of creation, which connotes the separation and the superiority of good over evil: "Behold, the devil is about to throw some of you into prison, that you may be tested, and for ten days you will have tribulation. Be faithful unto death, and I will give you the crown of life." (Revelation 2:10).

Message to Pergamos:

> Criticism is directed in this particular instance against false philosophies which are quite in contrast to the "dry land" state of thought of the *third* day of creation. Moreover, the "white stone" promised to those that overcome such false doctrines symbolizes the firm foundation offered by the "dry land" state.

Message to Thyatira:

> The false teachings and principles attacked here may be compared with the *fourth* day of creation, stressing that God is divine System and Principle which, eventually, must rule the world: "He who conquers and who keeps my works until the end, I will give him power over the nations, and he shall rule them with a rod of iron ... ." (Revelation 2:26-28). The "morning star" also promised is a further symbol of divine System or Principle.

Message to Sardis:

> In this message we have criticism directed against many who are "dead" spiritually – quite the opposite of the *fifth* day of creation,

which is characterized by the abundance and multiplication of Life: "I know your works; you have the name of being alive, and you are dead." (Revelation 3:1).

Message to Philadelphia:

Since this church had made an effort to utilize its limited potentialities, it is promised dominion over those whose thought is still darkened: "Behold, I will make them come and bow down before your feet." (Revelation 3:9). In addition, it promises intimate communion and identification with God: "He who conquers, I will make him a pillar in the temple of my God; never shall he go out of it, and I will write on him the name of my God." (Revelation 3:12). This is precisely the essence of the *sixth* day of creation – dominion and recognition of the immortal nature of both God *and* man!

Message to Laodicea:

This church is accused of thinking itself, without justification, to be in harmony with the sense of the *seventh* day of creation which is characterized by a state of Perfection: "For you say, 'I am rich, I have prospered, and I need nothing; not knowing that you are wretched, pitiable, poor, blind, and naked.'" (Revelation 3:17).

The Seven Seals, Trumpets and Vials:

Since we cannot hope to go into the many instructive details of the *seven* seals, the *seven* trumpets and the *seven* vials, it must suffice to say that they are introduced into the text to symbolize a s*evenfold destruction of false beliefs.* For instance, let us take the sounding of the fourth trumpet as an example: "The fourth angel blew his trumpet, and a *third* of the sun was struck, and a *third* of the moon, and a *third* of the stars, so that a *third* of their light was darkened; a *third* of the day was kept from shining, and likewise a *third* of the night." (Revelation 8:12). Understanding that it is false mortal beliefs which are being destroyed, and as a result recognizing the usage of the symbol "three" in its opposite sense, it will be seen that there is an attempt to convey the message that eventually all false systems and principles are bound to fall. (The Sun, Moon and stars of the fourth day of creation are used as a symbol of System or Principle.)

The Red Dragon and the Woman:

The episode in the twelfth chapter of Revelation, concerning a pregnant woman persecuted by a "great red dragon," provides a typical example of the symbolism used to illustrate birth of the *Christ-idea* into the world. The woman wearing a "crown of *twelve* stars" is seen to be introduced as a symbol of the authority and universal extent of divine Principle; and the child which she was to bring forth is the *Christ-idea* being born into the hearts of mankind. Opposing such an event we have the "great red dragon," which is a symbol of a mental state of darkness and ignorance. (The *seven* heads and *seven* crowns together symbolize imperfection and the opposite of insight into the spiritual meaning of the seven days of creation; the *ten* horns are a counterfeit of the Ten Commandments; and the *third* part of the stars of heaven which its tail cast to earth is a symbol of a total lack of insight into the Truth.)

However, as formidable as the "great red dragon" is depicted, and despite the efforts of imperfect mortal minds to prevent the birth of the *Christ-idea,* it is recorded that "the earth came to the help of the woman, and the earth opened its mouth and swallowed the river which the dragon had poured from his mouth." (Revelation 12:16). The intended moral, of course, embraces the prediction that eventually the world will turn against the "great red dragon" state of mind and come to accept the divine System of Principle we call God.

The "New Jerusalem:"

The "new Jerusalem" or city "foursquare" of Revelation, chapter 21 refers not to a physical structure, of course, but rather to the establishment of a new sense of government based upon Truth. The "foursquare" aspect is symbolic of much sought after Truth; while the *twelve* gates of the city and the *twelve* precious types of jewels which adorned the very "foundations of the wall of the city" together symbolize that one may enter into Truth only through a demonstration of divine Principles. Upon gaining admission we will indeed be blessed with a realistic perspective and possess something truly precious.

The fact that "I saw no temple in the city, for its temple is the Lord God the Almighty and the Lamb" (Revelation 21:22) conveys the important message that organized religion and the idea of a "place of worship" must one day pass away, since the *Christ-idea* – as expressed through the medium of an individual's conscience – will be seen to afford the most successful route to Truth. Imparting a quite similar message is the statement: "And the city has no need of the sun or moon to shine upon it, for the glory of God is its light, and its lamp is the Lamb." (Revelation 21:23). For with universal achievement of a higher perspective there will no longer be any need to express matters in symbolic terms or to rely upon some external source of light, since the light of conscience will be both clear and adequate.

Words of Lasting Wisdom

As in the past, the New Testament continues to be misunderstood,* suffering much at the hands of skeptics and fanatics alike. But while it has been distorted and abused by so-called religious "authorities" masquerading under false pretenses, it nevertheless contains a wealth of constructive philosophy. The following are but a few of the many rich and enlightening passages taken from this unique collection of literature, and which all of us would do well to ponder:

"And you will know the truth, and the truth will make you free." (John 8:32)

"For where your treasure is, there will your heart be also." (Matthew 6:21)

"But seek first his kingdom and his righteousness, and all these things shall be yours as well." (Matthew 6:33)

"Truly, I say to you, it will be hard for a rich man to enter the kingdom of heaven." (Matthew 19:23)

"For what does it profit a man, to gain the whole world and forfeit his life?" (Mark 8:36)

"And as you wish that men would do to you, do so to them." (Luke 6:31)

"Let your light so shine before man, that they may see your good works and give glory to your Father who is in heaven." (Matthew 5:16)

---

*   An informative account of the rise of the Christ myth may be found in a book entitled *Who Wrote the New Testament? (The Making of the Christian Myth)*, by Professor Burton L. Mack (published in 1995 by HarperCollins; ISBN 0-06-065517-8).

9

Christianity and the Dead Sea Scrolls

Discovery of the Scrolls – Dating of the Scrolls – Contents of the Scrolls – Christian Origins – Suppression of the Scrolls

The issue of Christian origins is not a subject that our churches generally care to discuss. They prefer to let their flocks live a rather sheltered life, oblivious to important new evidence and to the impact of scholarly research over the past hundred years or so. In particular, they have endeavored to pretend that discovery of the Dead Sea Scrolls, during the middle of the 20th century, has posed little or no threat to traditional Christian beliefs. This could hardly be further from the truth.

Discovery of the Scrolls

Roughly 20 miles east of Jerusalem lies the northern tip of the Dead Sea, a land-locked body of exceedingly salty water extending southward almost 50 miles and with an average width of close to 10 miles. Constituting the lowest sizable body of water in the world, its surface is situated some 1,290 feet below sea level. Although less than 20 feet deep at the south end, its greatest depth is toward the northern edge where it drops to about 1,300 feet. Some six times saltier than the ocean, it is surrounded for the most part by rugged terrain and a series of steep cliffs containing numerous fissures and caves. It was from just such caves – near the northwest shore – that a wealth of ancient documents first surfaced in 1947, and findings continued over the course of the early 1950s.

CHRISTIANITY AND THE DEAD SEA SCROLLS

At the extreme northwest corner of the Dead Sea, little over a mile from this unique body of water, is to be found the ruins of what was once a combination religious housing settlement and fortress known as Khirbet Qumran. Dating back to the first half of the 2nd century BC, it was badly damaged and abandoned during the reign of Herod the Great, from 37 to 4 BC. Eventually rebuilt, it stood until finally destroyed for good by the Romans in 68 AD. Identified by historians as being occupied by a Jewish branch sect known as the Essenes, it was from just such a group that a vital key would be found to link the Dead Sea Scrolls to Christian origins.

The Essenes evidently populated many local communities, including nearby Jerusalem which was home to a large following. Over the years their once ascetic, reclusive and pacifist membership became deeply infiltrated with a number of religious Zealots – men "zealous for the Law" of strict Judaism and antagonistic toward their Roman invaders, whom they regarded as an enemy to be expelled by whatever means were necessary. Although called Essenes by several ancient historians, inhabitants of the Qumran community often referred to themselves as "followers of the Way," and ran their own social order upon the basis of a communistic economy – in which all wealth and possessions were shared and held in common. It was truly a distinctive philosophy which separated this group from the mainstream of Jewish society.

The actual story of the Dead Sea Scrolls began early in 1947, when a Bedouin shepherd boy noticed a small opening in the face of a cliff not far from the Qumran ruins. Crawling inside, he soon discovered a number of earthenware jars. Subsequent investigation would reveal at least forty jars, some broken and others still intact with sealed lids. Not realizing the intrinsic worth of the find, Bedouin tribesmen removed a quantity of leather rolls (wrapped in decaying linen) from the jars. While it is not known for certain just how many scrolls were taken, a total of seven (more or less intact) documents eventually found their way into the hands of a number of responsible authorities, along with fragments of almost two dozen others.

A brief description of these seven principal scrolls follows:

1) *St. Mark's Isaiah Scroll* is a well-preserved and complete copy of the Old Testament Book of Isaiah. Measuring some 24 feet in length and roughly 12 inches wide, it is the oldest intact copy of this ancient manuscript.

2) *Hebrew University Isaiah Scroll,* with most of the first half missing, is likewise a very old copy of this popular Old Testament docu-

ment. Differing only a little in text from the rendition found in the Bible today, its leather is much deteriorated.

3) *Habakkuk Commentary* is a commentary of the Old Testament book with new text applicable to events and ideas which were current at the time of the commentator. Measuring about 5 feet in length and just under 6 inches in width, it has a few pieces missing.

4) *Community Rule,* or "The Manual of Discipline," deals with rules of the Community and is also "sectarian" in nature (i.e., text dealing with contemporary ideas and issues). Roughly 6 feet long and about 10 inches wide (with its two separated leather sections joined), it also has a few pieces missing.

5) *The War Scroll,* or "The War of the Sons of Light with the Sons of Darkness," is a well-preserved apocalyptic and eschatological work which is similar in many respects to the Book of Revelation. This document measures about 9 feet by 6 inches.

6) *Thanksgiving Psalms* consists of a collection of psalms and parts of psalms not unlike those of the Old Testament. Considered as a form of literature, in the eyes of many scholars they are deemed to be slightly inferior.

7) *Genesis Apocryphon,* or "Aramaic Scroll," is written in the Aramaic language instead of in Hebrew. It contains chapters from the Book of Genesis, with somewhat expanded and embellished text.

Encouraged by the wealth of ancient documents found in what came to be known as Cave 1, it was inevitable that serious efforts would follow to explore every cave and crevasse in the vicinity of Qumran. In due course, a number of other caves, containing a large corpus of scrolls, were discovered. They were to set the stage for a scenario of intrigue and scholarly debate.

One particular find, known as the *Copper Scroll,* soon proved to be highly fascinating. Uncovered in March of 1952, in a site since designated as Cave 3, were two segments of a single scroll of rolled copper. With writing punched into the metal, it was obviously made to last and to withstand the ravages of time. Too brittle to be readily opened, it was not until 1956 that both halves had been sliced open and deciphered. It told of a vast treasure (mostly gold and silver) scattered over no less than 64 locations, with the focal point being the city of Jerusalem. Almost certainly, it had to refer to concealment of the Temple treasure in the face of an impending invasion from a large Roman army, which was dispatched in response to the Jewish rebellion of 66 AD. Unfortunately, the locations listed in the text have been rendered essentially meaningless with the passage of so much time, as reference points had long since faded into obscurity. To date, all attempts to find any portion of this huge inventory of wealth have ended in failure. Conceivably, the frequent usage of "symbolic numerals" leads one to suspect that many quantities have been inflated – especially since the amounts appear to be far beyond what is likely to have been accumulated. (Examples of this symbolism may be cited as follows: "Beneath the steps that go to the east 40 cubits, a chest of silver; its totality of weight is 17 talents" ... "going down to the east 3 cubits up from the bottom, 40 talents of silver," etc.) If nothing more, it shows that the author of this scroll possessed insight into the unique system of numerical symbolism.

But the real treasure-trove lay in the deep recesses of Cave 4, discovered in September of 1952, as it contained sizable portions of more than 500 separate documents and over 15,000 small fragments. Incredibly, this cache was situated within a scant 50 feet of the Qumran ruins! Without a doubt, it was a regional depository for the community's extensive library of biblical manuscripts, hidden for safekeeping from an advancing Roman army. (It is conceivable, of course, that many of the cave documents came from Jerusalem.) Copies of every Old Testament work, with the sole exception of the Book of Esther, were uncovered. (It would seem that the Esther "fairy tale" story, depicting a Jewess who is compelled to marry and become the queen of a pagan king, did not find favor with the Qumran leadership who were highly nationalistic in outlook.) While numerous duplicate copies of much popular literature characterized the find, there were also many sectarian works and previously unknown manuscripts. It was in these latter scrolls – dealing with contemporary issues – that lay the fundamental worth of the discovery, which soon led to a truly unprincipled campaign of deception and concealment.*

---

* A most informative account of this deception – along with much valuable insight into the latest scholarly research relating to the Scrolls and Christian origins – may be found in an excellent book entitled *The Dead Sea Scrolls Deception* by Michael Baigent and Richard Leigh (published in 1991 by Summit Books; ISBN 0-671-73454-7).

In order to deal effectively with such an enormous corpus of material, it was clear that the talents of many experts would be required to decipher and publish the 800 or so separate documents which had surfaced by the end of 1952. Accordingly, a team was assembled and actively engaged in this monumental task by the summer of 1953. Inexplicably, it excluded all Jewish scholars, being comprised chiefly of Roman Catholic priests under the leadership of Father Roland de Vaux – a highly opinionated and bigoted Dominican priest. Needless to say, from the team's very inception temptation to withhold or publish a biased interpretation – in favor of dogma inherent in traditional Christian beliefs – was bound to be quite irresistible!

Dating of the Scrolls

In the interest of "preserving the faith," it soon became clear to de Vaux that it would be dangerous to link the sectarian scrolls to the early Christian era. (Although none of the canonical Old Testament books posed a threat, there was a large corpus of sectarian literature which dealt with contemporary issues and, as such, they could very well serve to undermine traditional beliefs.) In fact, the further one could distance them from the time of Jesus the better for orthodox theology. Thus it came to be that dating the Dead Sea Scrolls, or of associating them in any way with Christian origins, was crucial to the survival of church doctrine.

A number of diverse procedures are employed by scholars to date ancient manuscripts. In this instance, the following methods deserve discussion:

1) Carbon-14 testing;

2) Artifacts;

3) Paleography;

4) History; and

5) Internal evidence.

Radiocarbon dating techniques, first developed in 1947, offer a promising means of determining the age of an organic specimen. Based upon knowledge that the radioactive isotope carbon-14 (produced in the atmosphere by cosmic rays and incorporated into all forms of living matter) has a half-life of about 5,730 years, before decaying into common carbon-12, it becomes feasible to derive an age by the simple expedient of assessing respective ratios of the two isotopes. Improved considerably over the years, modern AMS (Accelerator Mass Spectroscopy) testing is far more accurate and wastes less material in the process. Carbon-14 dating is able to give reliable ages of organic material in excess of 50,000 years, although accuracy does fall off somewhat with age.

Shortly after their discovery, a piece of linen wrapping from the Scrolls was subjected to such a test and given a date of 33 AD – but with a plus or minus correction factors of some 200 years. Unfortunately, while this initial test promptly ruled out a medieval origin for them, it left too great a margin of uncertainty to settle the issue one way or the other. What really tends to boggle the mind is that, for more than 40 years, no additional and refined carbon-14 tests were ever performed upon these ancient documents!

Excavation of the Qumran ruins has left little doubt that it was once a dual purpose fortress and communal center with a library and, perhaps, a small scriptorium. Literally hundreds of coins were found, dating from early 1st century BC to the Jewish revolt of 132 to 135 AD under Simeon bar Kochba. An abrupt decline of coins from 68 AD coincides well with several historical records indicating destruction of the settlement by the Roman army at this time. The style and shape of the pottery jars, which contained the Scrolls, is also deemed by a number of experts to be of 1st century AD antiquity.

While far from an exact science, the paleographer can often date writing by the subtle manner in which alphabet characters and words change with the passage of time. Upon examining various samples, it was determined that there was indeed indication of a degree of some evolutionary change consistent with a range of dates extending from 1st century BC to at least mid-1st century AD.

History, as a means of dating, may be invoked in order to supply a reason for hiding the Scrolls in the depths of barely accessible caves. In this regard, only three possibilities – all involving a Roman invasion – need be considered: Pompey in 63 BC; Vespasian in 66 AD; and Hadrian in 135 AD. (Actually, some documents pertaining to this latest era were found in caves more than 11 miles to the south of Qumran, including two letters signed by Zealot rebel leader Simeon bar Kochba. Relating solely to matters of the later rebellion, they are merely of academic interest.)

The very contents of the sectarian documents themselves afford a measure of rather conclusive proof as to which of the three scenarios is the most probable. Among a wealth of internal evidence which may be adduced, through an examination of the published Dead Sea Scrolls, are the following examples:

Copper Scroll:

This scroll was almost certainly deposited for safe-keeping (in Cave 3) in the wake of the 66 AD revolt. Since there was no Jerusalem Temple in 135 AD, from which a treasure might be obtained to hide, this late date is thereby ruled out. Moreover, there are few (if any) scholars who would seriously entertain the idea that this scroll could date back to the time of Pompey.

Roman Invaders:

Both the "Habakkuk Commentary" and the "War Scroll" documents allude to a Roman invasion – but one whose generals are commanded by a "king" or "monarch." Until the year 27 BC Rome was still a republic without an emperor. Thus the 63 BC date must be ruled out.

Herodian King Marriages:

In the entire corpus of Old Testament literature there is no mention of a Jewish king being forbidden to marry his niece. In two Qumran sectarian documents ("Temple Scroll" and "Damascus Document") this is strictly prohibited – revealing it to be a 1st century AD concept directed at Herod! Since John the Baptist was executed for vigorously condemning the marriage of Herod Antipas to his niece, this clearly rules out all but the 66 AD date.

Wicked Priest, Liar and Teacher of Righteousness:

These three unnamed characters appear together quite often in a number of the Qumran sectarian writings, most notably in the "Habakkuk Commentary" and the "Damascus Document." As we shall presently see, all three can be identified with personages contemporary with Jesus, and who lived for a number of decades after his death. Accordingly, such documents may be dated around mid-1st century AD.

It will therefore be concluded that, beyond any reasonable doubt, all of the aforementioned sectarian documents are editions which were written within a limited 1st century AD time frame – most likely during the life of Jesus, and extending for some years until the destruction of Khirbet Qumran in 68 AD.

Contents of the Scrolls

Of the published sectarian Dead Sea Scrolls, which is the only type of literature capable of imparting insight into Christian origins, the following documents are deserving of special comment and elucidation:

War Scroll:

Copies of this document were found in Caves 1 and 4, both of which are situated in the immediate vicinity of the Qumran ruins. This work was almost certainly written in response to an impending "Kittim" (Roman) invasion of their community. Depicting a symbolic war between the "Sons of Light" (themselves) and their adversary, the "Sons of Darkness" (the Romans), it expresses the notion of eventual triumph if their members are "zealous for the Law."

Instead of a semi-monastic atmosphere characterizing Qumran – as orthodox theology had once supposed – it is now evident that this community had been infiltrated by large numbers of Zealots who were prepared to fight with great fervor. Indeed, since texts identical to several Qumran sectarian scrolls were found at Masada (a cliff fortress some 30 miles to the south, which was finally subdued by the Romans in 74 AD), it is clear that the two groups were essentially one and the same!

Community Rule:

One of the original finds in Cave 1, it deals primarily with rules and moral code within the Qumran community which, it may be stressed, was not restricted to just the inhabitants of this one physical structure. On the contrary, there were probably many more "followers of the Way" (Acts 9:2) that lived in Jerusalem and other neighboring settlements. Glossing over assorted trivia, there are two rather remarkable parallels with New Testament writings which deserve mention.

The first and most striking observation involves a "communistic" society, in which it is stressed that all wealth and assets are to be shared equally by members of their community. A fundamental statute, within the text of this scroll, is the requirement of new members to "bring all their knowledge, powers and possessions into the Community." Another passage states that "his property shall be merged and he shall offer his counsel and judgment to the Community." This is precisely the gist of Acts 2:44-45, in which it is stated: "And all who believed were together and had all things in common; and they sold their possessions and goods and distributed them to all, as any had need." One can hardly escape the conviction that many of the original followers of Jesus (only later to be called Christians) were intimately associated with the Qumran community – presumably, for the simple reason that a branch of this society constituted the very nucleus of the early church!

Yet another link between the New Testament and the "Community Rule" is the way in which they utilize similar phrases and expound similar concepts. Jesus' famous "Sermon on the Mount," for example, is common to both bodies of literature. (This fact will be seen to present somewhat of a dilemma for Christian theologians who, in a desperate move to distance Qumran from Jesus by considering the Scrolls to be pre-Christian, would then have to concede that Jesus was merely echoing an earlier work.) Again, emphasizing their great "zeal for the Law," this document expressly states that anyone who "transgresses one word of the Law of Moses, on any point whatever, shall be expelled." This is essentially what Jesus is reported to have said: "Think not that I have come to abolish the law and the prophets; I have come not to abolish them but to fulfill them. For I say to you, till heaven and earth pass away, not an iota, not a dot, will pass from the law until all is accomplished." (Matthew 5:17-18). Once more, we are compelled to acknowledge a definite connection between Jesus and the Qumran community.

Temple Scroll:

Believed by some to have been discovered in Cave 11 by Bedouin tribesmen, this scroll eventually found its way into the hands of a local antiquities dealer by the name of Kando. Six years later it was turned over to Israeli authorities and finally published in 1977. Appearing on the surface to deal with little more than trivia relating to furnishings, laws and rituals involving the Temple in Jerusalem, it does provide unexpected insight into certain facets of 1st century AD history and a link to another more valuable work.

The "Temple Scroll" reveals that the Qumran sect was not nearly as divorced from the mainstream of Jewish religious life as once thought. To them the Temple was still their center of worship and its ruling theocracy was still very much their concern. In fact, it was the leaders of the "followers of the Way" who deemed themselves to be the true priests of the Temple, even if not designated as such by any Roman puppet king. Feeling justified by authority, it was merely a short step to write (or edit) scriptural text to reflect their own ostensibly superior views. Thus it came to be that text was composed forbidding a Jewish king to marry his niece. (Looked upon by the Qumran leadership as a foreign king, and not one "zealous for the Law," they either modified earlier text or composed their own in order to cast Herod Antipas in a bad light.) In any event, they performed the useful function of dating this scroll to within a relatively narrow time span.

As it turned out, this dislike of King Herod served to clarify a long-standing mystery regarding another biblical writing known as the "Damascus Document." Discovered in 1896, in the loft of an old synagogue in Cairo, scholars had hitherto been unable to date it or to place it into context with other literature. However, with knowledge of the "Temple Scroll," they could see that it belonged to the Dead Sea library of documents. This connection was now firmly established by the very fact that, of the entire corpus of earlier biblical writings, only the "Damascus Document" and the newly-found "Temple Scroll" spoke of a prohibition against a king marrying his niece!

Damascus Document:

Portions of eight copies of the "Damascus Document" were found in Cave 4 at Qumran; while parts of the ninth were discovered in Cave 5. An incomplete tenth copy was also uncovered in Cave 6, indicating that it was a popular work. Known to scholars for more than half a century, this document was given its name by reason of a reference to a place in the wilderness called "Damascus," where an unspecified "Teacher of Righteousness" is described as leader of a Jewish sect that remained "true to the Law."

What is so highly significant about the name of this place is that it is able to clear up a mystery with regard to Acts 9, of the New Testament, where Paul is reported to have been blinded for three days while on the road to "Damascus." Previously, it had been quite wrongly assumed that Damascus in Syria was the city in question. But this had always posed a problem, since an arrest warrant issued in Jerusalem would have been useless in such a foreign country! Thus there can no longer be any doubt that this communal residence (or fortress) was originally called "Damascus" by its inhabitants. (As we shall presently see, this information has all the aspects of a "Rosetta stone" – one capable of revealing the truth about crucial events concerning the apostle Paul and Christian origins.)

Habakkuk Commentary:

This scroll was one of the first to be found in Cave 1 at Qumran. In conjunction with a number of other documents describing the same trio of mysterious characters, the "Habakkuk Commentary" poses a very real threat to traditional Christian beliefs. There is repetitive mention of a conflict involving a "Teacher of Righteousness," a "Liar" and a "Wicked Priest." Who are they? What role do they play in the scheme of events just prior to the Jewish uprising of 66 AD? Due in no small measure to the research of Professor Robert Eisenman (Chairman of the Department of Religious Studies at California State University in Long Beach), it is now possible to finally resolve such questions.

Although initially there was a strong temptation for orthodox theology to visualize Jesus as being the "Teacher of Righteousness," this idea is beset by a serious flaw. In short, there is not the slightest hint – in all the Dead Sea Scrolls – that their leader was ever considered to be divine by his followers! If those in immediate contact with Jesus did not look upon him as the "Son of God," then how could Christians today be expected to uphold such a belief? As a result, this notion had to be abandoned and the problem was left unanswered by Christian theologians.

The key to identifying the three aforementioned participants mentioned in the conflict is to be found in the "Liar," who is described as being a traitor from *within* – one who had been taken in by the Qumran community, accepted as an accredited member after an initiation period, who subsequently defects. Recalling that Paul of Tarsus was greatly changed after his experience while on the road to "Damascus" (Qumran), it is most logical to picture him as the "Liar." He does, in fact, meet all of the requirements, and he was in the right place at the right time. But, one might ask, in what sense did he defect? Quite simply, he was the first to preach that Jesus was anything more than a human being – that he was, indeed, an incarnation of God!

The identities of the other two persons, the "Wicked Priest" and the "Teacher of Righteousness," promptly fall into place. The high priest Ananas was the Temple priest at this time and is reported to have been an adversary of James – an early pre-Christian leader – and therefore a "Wicked Priest." By association, it seems most likely that James is the unnamed "Teacher of Righteousness." (Conceivably, this esteemed title could refer to any current leader of the Qumran community. In any event, James would definitely appear to be the one so described in this instance.)

Armed with the identities of the quarreling trio mentioned in such documents as the "Habakkuk Commentary," we are now in a position to proffer a truly revolutionary picture of Christian origins.

Christian Origins

As we have seen, canonical New Testament literature is not a very reliable guide when it comes to dealing with actual fact, since it is heavily biased in favor of imparting a religious message. Thus, in order to place the issue of Christian origins into true perspective, it is desirable to begin with what is known as historical fact, and then to extrapolate a probable scenario from the new evidence which is implicit in the Dead Sea Scrolls. To this end we shall commence with the lives of Jesus, James and Paul – the three principals who were involved initially in what later evolved into our present-day Christianity.

The Personage of Jesus:

Most likely born between 6 and 4 BC, somewhere in the general vicinity of Jerusalem, Jesus appears to have been well-versed in all aspects of biblical scripture. Highly antagonistic toward both the Sadducees and Pharisees, we are left with only the neighboring Qumran community as a possible source for his later education. Indeed, with the evidence strongly weighted in this direction, it can hardly be doubted that Jesus must have spent considerable time studying and conversing with members of this particular Jewish sect. Eventually, convinced that he had learned all that they could teach him, he then departed and began to preach on his own.

It is likely much more than a coincidence that at least some (if not most) of Jesus' original followers were recruited from among the Qumran membership. In particular, it is now little short of obvious that John the Baptist must have been intimately entwined with the Qumran sect – especially since he was so zealous in defending Qumran policies (e.g., the Herodian king/niece marriage affair) that he was executed as a result. By reason of his close association with John the Baptist, it must be concluded that Jesus was of the same school, so to speak.

Unfortunately, the infusion of Zealot elements into the Qumran sect was to prove the undoing of Jesus. That he was indifferent or even sympathetic to their cause is attested to by the mention that "Simon who was called the Zealot" (Luke 6:15) *was actually one of Jesus' chosen apostles!* Also testifying to the rebellious company that he kept was the incident in the garden of Gethsemane, where Jesus' disciples are described as being an armed band. (Peter is reported to have cut off the right ear of the servant of the high priest.) Another occasion in which violence was displayed is the episode of Jesus overturning the tables of the money-changers in the Temple (Matthew 21:12). Moreover, statements like "Do not think that I have come to bring peace on earth; I have not come to bring peace, but a sword" (Matthew 10:34) can only make one think of the militant Zealots. Yet a further indication of his involvement with this "zealous for the Law" sect is the fact that the Romans saw fit to crucify him – a punishment employed exclusively against political and rebellious offenders. Either Jesus was framed, or he was himself a member of this extremist group!

It is, therefore, to be concluded that there are two sides to the personage of Jesus. On the one hand, he would seem to have been an exceptionally wise and enlightened philosopher for his era. By way of contrast, he might well have become unwisely involved with a group displaying a little too much "zeal for the Law." (This association is clearly attested to in Acts 5:35-39, where Jesus' disciples are compared to earlier bands of militant Zealots.)

The Personage of James:

Mentioned in the Bible as a brother of Jesus (**Gal**.1:19), the character of James has been quietly pushed into the background by church theologians. (Although it is just conceivable that the term "brother" is a metaphor for a fellow member of the Qumran society, it still remains a fact that he was closely associated with Jesus.) Indeed, much of the early literature relating to James is rather mysteriously excluded from the New Testament canon. Why? This is even more surprising when it is considered that James is described as head of the early church in Jerusalem – instead of Peter, who is supposed to be the "rock" upon whom Jesus would build his church (Matthew 16:18). The answer is one which completely undermines the very fabric of orthodox Christian beliefs: James was a staunch proponent and upholder of the "Law of Moses," *and as such would have nothing to do with the wild idea of considering Jesus to have been an incarnation of God!*

This interpretation explains the pronounced rift between Paul and James, since the former was actively preaching a story about Jesus' resurrection and divine nature. Unable to tolerate such an obvious fabrication, and as the "Teacher of Righteousness" head of the Qumran/Jerusalem sect, James is forced to object and to publicly admonish Paul. In turn, Paul's repeated assertions gets him into serious trouble with James' zealous "followers of the Way," and he is only saved from death at their hands by Roman soldiers (Acts 21:32).

James' second adversary was the Temple high priest Ananas (depicted as the "Wicked Priest" in several documents of the Dead Sea Scrolls). Considered by James to be a puppet of Roman authorities, Ananas was reviled by Qumran "followers of the Way." Despised in turn by the "Wicked Priest," James was looked upon as an unwelcome rival of the Temple establishment. Moreover, by reason of his Zealot connections, James was a threat to Temple security and a feared Roman reprisal. It was an atmosphere in which something bad was bound to happen.

When violence finally broke out the first major casualty would be James, who was severely beaten, eventually perishing (around 62 AD) at the hands of Ananas' hostile supporters. Incensed by this attack against their leader, the Zealot faction within the Qumran community subsequently managed to exact a measure of vengeance by assassinating the "Wicked Priest." Not long thereafter, all of Jerusalem exploded into violence, culminating in a general uprising and the Jewish armed revolt of 66 AD.

The Personage of Paul:

Born in the city of Tarsus (now in Turkey), Paul was of Jewish heritage with Roman citizenship. Well-educated, he was fluent in several languages. Never having met Jesus, he enters the biblical scene as an official entrusted with the task of persecuting members of the early church. At this time little more than another radical Jewish sect, infiltrated with Zealots, there was no thought of the Qumran group ever becoming the nucleus of a new and dominating religion. Merely a simple expedition to "Damascus" (Qumran) in order to arrest certain ringleaders of the "followers of the

Way" – men who were suspected of being militant Zealots – this seemingly insignificant trip would serve to drastically change the course of history.

According to the text in Acts, chapter 9, Paul (called Saul, at this point in time) is in charge of a band of armed men en route to Qumran when a "light from heaven" supposedly knocks him to the ground. Whether it be pangs of conscience for previous misdeeds, or plain sunstroke, he is reportedly stricken blind for *three* days. When insight into the truth (the symbolic meaning of the numeral three) is imparted to him by a member of the Qumran community, he is completely changed. Allegedly spending three years at Qumran (Galatians 1:17-18), Paul is converted and accepted into their society as a legitimate member. He then departs to join James in Jerusalem, ostensibly most anxious to redeem himself by now preaching the sect's "zeal for the Law."

Evidently encountering a measure of distrust from some of James' associates (understandable in view of past experience), it is not long before Paul is farmed out to more remote locations – clear out of the country, in fact. This turns out to be a big mistake. Thrust into contact with other pagan religions boasting of a variety of most impressive deities – including semi-gods with magical powers – it is only a matter of time until Paul succumbs to a strong temptation to embellish his account of Jesus in order to be more competitive. In due course, he finds himself making exaggerated claims and preaching a miraculous resurrection from the dead. The people just love his stories and he soon begins to make large numbers of converts. The ground is fertile and the promise of a heavenly life is exactly what the masses want to hear.

However, word of Paul's deification of Jesus eventually reaches James in Jerusalem. Paul is subsequently recalled and rebuked for his actions. A compromise of sorts is reached, to the effect that some aspects of the "Law of Moses" will be relaxed for non-Jews to join the church in the form of second-rate members. In return, Paul agrees to tone down his preaching and stick to Qumran doctrine. He then embarks upon what must be described as a rather extended and incredibly successful mission – the only problem is that his many converts believe in a Jesus/God deity!

Again recalled to Jerusalem by James some years later (about 58 AD), Paul foolishly thinks that he can still come to an equitable agreement with a church leadership that has by now little in common with Pauline philosophy. Confronted with the evidence, Paul is soon branded a traitor and a "Liar" by the Qumran community, in spite of the monetary contribution which he brings to the church from his recent converts. Ambushed by a mob of pious Qumran extremists, much less tolerant than James, Paul is barely saved from sure death by the intervention of Roman soldiers. Following a somewhat lengthy series of questionings, by various authorities, he is shipped off to Rome to face some kind of trial, where he fades out of the picture.

But the die is cast. In short order the post-Jesus church is superseded by a new and radical Christ-Jesus religion – one which very quickly spreads to surrounding nations, and which bears little resemblance to the philosophy expounded by its alleged founder!

An Unplanned Religion:

It must now be abundantly clear, to all who would seek truth in an unbiased manner, that present-day Christianity is built upon a framework of deception and lies. Its major premises are seen to have absolutely no basis in fact and can only serve to disgust those who would exercise the power of human logic. The great irony of it all is that the theological structure of this most peculiar Christ/Jesus religion was never intended by Jesus, who was simply trying to uphold the "Law of Moses."

Following in Paul's footsteps, the new religion was soon further embellished by a succession of deceived and over-zealous theologians, scriptural writers and editors. Thus it was that Old Testament ideas of a "Messiah," who would lead the Jewish race to world supremacy, was replaced by a New Testament theology embracing the wildest of claims. First, there was the notion that a mere human being was God! Then, incredible as it might seem, it was claimed that through faith and blood redemption man could be "saved" and pass to a "heaven" – completely bypassing the vital principle of free will and contrary to a cosmic requirement that destiny must lie in the hands of the individual!

To date, organized religions have relied upon blind faith and lack of knowledge in order to retain their flocks. Hopefully, with fresh insight into the Dead Sea Scrolls, it will not be long before nebulous faith gives way to the firm knowledge that is capable of opening new horizons.

Suppression of the Scrolls

Confronted by the great wealth of material discovered in Cave 4, it was apparent that some sort of master plan would be needed to decipher and assess the huge find. Accordingly, the idea of forming an International

Team of scholars and linguistic experts was soon conceived. With headquarters at the Rockefeller Museum, located in the Jordanian sector of East Jerusalem (where Israelis were virtually excluded until the war of 1967), this team commenced work in 1953. Unfortunately, the membership of this team left much to be desired, and promptly set the stage for subsequent deceit and scandal.

The head of this historic group, and editor-in-chief of all the published documents until his death in 1971, was Father Roland de Vaux. This was just about the worst possible choice that could have been made, as he was a highly zealous Roman Catholic priest who was extremely bigoted against Jew and non-Christian alike. He was also an appointed consultant (1955) to the Pontifical Biblical Commission, an ecclesiastical body whose sole purpose is to "preserve the faith" – an office which may be looked upon as our modern-day successor of the medieval Grand Inquisitor! Already, there existed both motive and opportunity to cloud the truth.

Appointed under de Vaux's authority were the following original team members:

Professor Frank Cross:
> From the Albright Institute in Jerusalem, and originally associated with McCormick Theological Seminary in Chicago.

Monsignor Patrick Shehan:
> Also from the United States and, at the time, the director of the Albright Institute and an esteemed Roman Catholic priest/scholar.

Father Jean Starcky:
> An expert in Aramaic linguistics, and a Roman Catholic priest/scholar from France.

Father Josef Milik:
> Originally a Polish Roman Catholic priest who lived in France, he was a close associate and confidant of de Vaux – so close, in fact, that he would later be the recipient of some of the most sensitive sectarian material.

Dr. Claus-Hunno Hunzinger:
> A German expert who would shortly leave the team and be replaced by *Father Maurice Baillet,* another Roman Catholic priest from France.

John M. Allegro:
> The only team member without specific Christian affiliation; a highly qualified British philologist with a number of publications in prestigious academic journals to his credit. (He would eventually be phased out of the team and replaced by *Professor John Strugnell*, a Protestant theologian later converted to the Roman Catholic faith and a staunch supporter of Christian dogma.)

Incredibly, no Jewish scholars were included as members of what was supposed to be an "International Team" – in spite of the fact that many were well-qualified for the task of translating the writings of their own ancestors. But such was the bigotry which prevailed at this crucial moment of Dead Sea Scroll history. This inexcusable bias was to continue for almost four decades, during which time the leadership of the team would pass from *Father de Vaux* to his chosen successor, *Father Pierre Benoit*, in 1971. Upon Benoit's death in 1987, it passed to *John Strugnell*, so that the theme of supporting Christian dogma would be preserved. Moreover, not only was leadership inherited, but even the rights and all unpublished translations were bequeathed to designated successors. In effect, a small clique of religious bigots had somehow managed to hijack for themselves a priceless legacy which rightly belonged to all of humanity!

From the very beginning, it was deemed essential to distance the Scrolls from Jesus and Christian origins. To this end, de Vaux and his companions in deception strove diligently to withhold evidence that certain of these documents were written shortly before the Roman/Jewish war of 66 AD – the war which necessitated the concealment of the Scrolls in the dark recesses of the Qumran caves. Adopting a campaign of lame excuses and unprincipled delaying tactics, they were successful in suppressing full knowledge of their contents for many years. Fortunately, there was one member of the team who was not willing to go along with their nefarious plan.

Of the International Team's membership, only John Allegro was really free to examine the Scrolls in an impartial manner. Needless to say, a difference of opinion was bound to arise; it was just a question of when matters would come to a head. By 1956, Allegro had managed to get into serious trouble with de Vaux and his minions. Given access to less and less material of any consequence, it became clear to him that there was intention to phase him out. However, before he died in 1988 he had succeeded in stirring up quite a storm. Publishing a number of informative and highly controversial books, he brought to public attention the International Team's many inordinate delays and exposed their suppression of all sensitive documents. It is reported that in early 1959 he wrote to Awni Dajani, Jordanian curator of the Rockefeller Museum, expressing his fear that if any scrolls were to turn up which affected Roman Catholic dogma, the world would never see them. He was convinced that de Vaux was likely to send such objectionable material to the Vatican, where it would almost certainly be hidden or destroyed.

Regrettably, Allegro's courageous stand did not result in any immediate or substantial change of policy. Scrolls which should have been translated and made public in several years remained unpublished after several decades. Updated carbon-14 tests were yet to be made long after they were perfected to the point where they could have been most revealing. Photos of unpublished text were not available to impartial scholars until the early 1990s, and only then after a great public outcry and long overdue Israeli intervention, which finally resulted in the ouster of Strugnell as head of the team in late 1990.

But an interesting development of sorts is a story concerning Father Josef Milik, initially a close associate and right-hand man of de Vaux. Given a "lion's share" of the most important sectarian documents, Milik returned to Paris with photos of this material – which he alone had been entrusted to decipher – prior to the 1967 Arab/Israeli war. He eventually left the priesthood and became somewhat of a recluse. One is tempted to wonder at this unexpected and drastic change of attitude. What happened? Is it conceivable that, in studying this material, he found irrefutable evidence that completely undermined his Christian faith? If so, will the world ever learn the truth about such documents?

This very thought, of course, raises a disturbing possibility. In all the years of suppression, by a religious clique sworn to uphold and defend Christian dogma, just how many Dead Sea documents were actually destroyed on grounds that they posed a serious threat to the faith? Alas, we may never know, as no independent observers were ever permitted to catalog these ancient manuscripts for well over a generation after their discovery. Posterity will always wonder how, in this supposedly modern age, such a shameful situation could have been allowed to develop.

10

The Social Order

Society and the Individual – Society and Religion – The Economic System – In Search of a World Society

The nature of the social order under which we live is of utmost personal concern. It may mean existence in a state of freedom, security and happiness on the one hand; or on the other, it may subject us to poverty, injustice and virtual enslavement. But it is not enough to involve ourselves solely with the economic system which regulates our immediate lives. It is imperative to understand that a man is still a man and a woman is still a woman regardless of where or under what regime he or she lives and that a transgression against any individual is of concern to all who are worthy of membership in the human race.

In the past, the penalty for failure to comprehend and overcome religious and economic difficulties has led to wars that could have been avoided through knowledge. A similar failure in this present age of atomic and chemical/bacteriological warfare could well lead to a catastrophic annihilation of civilization itself. We have reached the point where modern means of destruction simply forbids us to condone future wars – where it is now most essential to achieve economic and social harmony among nations. It must, therefore, be perfectly clear that we can no longer afford to be indifferent to world affairs, and that there exists both a moral obligation and a dire necessity to acquire a realistic perspective with regard to such matters. A spirit of universal love and cooperation must prevail over petty nationalism and religious squabbling if there is to be any salvation for mankind.

Society and the Individual

The sanctity of the human soul has all too frequently been ignored by governments guilty of denying essential rights under the pretense of conserving freedom. What is so desperately needed, before it becomes too late, is a *World Constitution* effectively guaranteeing certain fundamental principles – namely, a set of rules respecting the dignity of man and backed by the might of a well-armed and dedicated military presence operating within the structure of a revitalized New United Nations.

The responsibilities of an acceptable social order involve, in the last analysis, a mixture of two opposing criteria. In contrast to a need for assurances of personal freedom and equality, there are also occasions when intervention is justified by the State. These rights and restrictions deserve a measure of discussion.

Human Rights:

The primary purpose of our existence is to evolve spiritually, through the process of overcoming an inherently selfish nature. *Any interference with this sacred obligation must be considered to be a violation of basic human rights!* Reduced to simple terms, the State must be charged with the preservation of four cardinal freedoms. These freedoms may be defined as follows:

2) Freedom to follow conscience:

> There are many ways in which governments have acted to the detriment of a soul in genuine pursuit of conscience. By far the most common invasion of privacy lies in the despicable practice of forcing an individual into military service. In the guise of upholding freedom, one is required to relinquish both freedom of mind and body. Moreover, at a command, he is expected to openly defy conscience by committing a transgression against his fellow man! Without a doubt, this can only be viewed as a flagrant example of enslavement. Not only is it criminal by its very nature, but submission provides the ammunition and the temptation for any unscrupulous leaders to wage warfare. It is a fact that if all who are drafted refused to enlist no major war would be possible! If nations were really sincere in their professed desire for peace, they would see the wisdom of outlawing compulsory military service. Under the auspices of a reformed New United Nations, a

The Social Order

worldwide ban on this outrageous custom could do much to prevent future large-scale wars and the devastating holocaust which must surely follow.

Linked closely to this deserving respect for conscience is freedom of speech and freedom from religious dogma. *Under no circumstance is the State justified in interfering with the free expression of ideas!* Indeed, if the prevailing social order is incapable of withstanding criticism, it is a sure sign that such a government is in dire need of revision.

2) *Freedom from want:*

While an unprincipled government cannot directly persecute conscience, it is to be regretted that it can have a most adverse influence upon the physical body housing the soul of man. With selfishness at the root of all evil, it is inevitable that spiritually immature leaders should seek to deprive others of a fair share of their rightful heritage. It is a fundamental law of simple arithmetic that whenever someone receives more than an average portion another must therefore go short. Since there is only so much wealth available for distribution, it may be deduced that in the making of every single millionaire many will be thrust toward (or into) poverty. Granted that some inequality with regard to material reward is desirable, or even necessary; but present extremes of wealth are quite unacceptable and constitute a prime cause of much of the world's unrest.

The obligations of a truly progressive State are quite clear with regard to the physical well-being of its citizens. While assuring a just share of material goods in return for honest labor, it is also duty-bound to provide abundant opportunities for gainful employment. Furthermore, in addition to acknowledging full responsibility for the support of the aged and infirm, it must display genuine concern for the health and welfare of all. In short, no individual can be allowed to suffer needless hardship for lack of sufficient funds.

3) *Freedom from discrimination:*

Discrimination may take many forms, with underlying causes ranging from outright greed to simple ignorance. No social order of worth can afford to condone injustice, whether it be prejudice of wealth, religion, politics, education, color, race or sex. In an atmosphere conducive to spiritual progress, every attempt will be made to eradicate inequality wherever it is seen to exist. The very foundation of a *Cosmic Perspective* is based upon developing an attitude of unselfishness – essentially, a feeling of universal love that would consider all humanity as equal.

Yet another aspect of discrimination is one in which governments think they have a divine right to admit or exclude inhabitants of our planet from crossing some nationalistic boundary. It is little short of imprisonment to prevent an individual from leaving a country, and an obvious violation of human rights. On the other hand, while there may be times when there is justification for a nation to impose some limitation upon permanent immigration, such a restriction should not apply to those whose presence would not impose a financial burden. No portion of our globe belongs to any one group or race, and the goal of an evolving world society is not to restrict, but to facilitate any exchange of culture.

4) *Freedom from aggression:*

Every individual has an inalienable right to be protected from all manner of undue persecution and bodily harm. It is a sad truth that there are governments which tend to regard life cheaply, often exhibiting little consideration for the welfare of its citizens. While such instances may, in part, be due to immaturity and selfishness in high places, it is also a fact that the very structure of the social order itself is largely to blame. Any society that permits serious inequalities of wealth and opportunity is prone to allow the wrong leaders to come to power, and thus cannot avoid fostering widespread resentment and frustration. It is surely no accident that the crime rate of a nation is related to its environmental conditions. By far, the best way to alleviate crime is to improve the social and economic system.

Human Restrictions:

In one respect, it may be said that present societies deliberately condone conditions potentially hazardous to the lives of its citizens. A most glaring example of this would be the incredibly lax gun control laws of some countries. (For a supposedly advanced society the United States must rank as one of the least intelligent in this regard.) Why any nation would purposely allow free access to hand guns is a great mystery, especially in view of the barely civilized attitude of some individuals and the usual quota of "borderline" mental cases that are forever walking around. Why put temptation so readily into the hands of such people?

Although many nations have recognized the great danger of drug abuse and are attempting to cope with the problem, for the sake of tax revenue and certain selfish business interests, they have neglected to target other insidious addictions that are detrimental to the health of virtually everyone. In monetary terms alone, the carnage of alcohol and tobacco manifestly exceeds what is collected by way of a license to play the fool. Ranging from personal injury or death at the hands of a drunken driver to a multitude of other atrocities, it is a fact that the alcoholic would be incapable of inflicting his mark upon society if that same society had never given its sanction! Similarly, if forbidden, an irresponsible and inconsiderate tobacco addict would not be allowed to endanger the health of innocent bystanders who are often forced to inhale their obnoxious and harmful smoke. Just as surely as though a man had taken a knife and stabbed his neighbor, so a thoughtless addict may achieve the same end result, albeit slower.

While it is clear that an individual is obliged to follow the dictates of conscience, *it is also evident that a line must be drawn should the actions of one threaten adversity to another!* This, of course, raises the issue of what to do with so-called criminals. In order to assess the problem of justice it is necessary, first of all, to fully understand the intrinsic nature of sin.

As far as the individual is concerned, sin may best be defined as a *failure to obey conscience.* And yet, as we are no doubt aware, our conscience can sometimes deceive us. It may tell us that a certain action is sinful, when actually there is nothing wrong; and right when it is really wrong. It is, in fact, a variable which is always prone to change as it evolves. Thus to follow our conscience is no positive assurance that what we are doing is absolutely right. Nevertheless, a violation of conscience – our own personal view of right and wrong – is definitely a sin. It is just as much a sin to violate conscience, even if it should be mistaken, as it is to deliberately reject the guidance of conscience when it is right. *A mistaken conscience calls for correction through the application of better light, but never by violation!*

The problem of dispensing true justice is one which mankind will never be able to solve completely. For one thing, we are incapable of properly evaluating the enormous effect that *environment* has upon the moulding of our characters. This point is of the utmost importance, since environment exerts an overpowering influence from which no one can escape. Indeed, were the truth about this factor really known, it would be interesting to see how many prominent citizens might react if subjected to the same circumstances as were faced by many whom we would judge so unfavorably. The results of such an experiment might well prove astounding: our penal institutions would likely be found to house a surprising number of quite unexpected guests!

Ever since the dawn of civilization it has been realized that some system is necessary to prevent a state of anarchy in which the strong are tempted to oppress the weak. Throughout the ages men have devised an almost endless variety of penalties and punishments (many of them outrageously unjust and inhuman) for what they considered to be acts of a criminal nature. While it is with a great sense of relief that we currently find ourselves living in a time that has put aside some of the more atrocious and warped of these concepts, we must not make the error of thinking that our present standards of justice are above reproach. On the contrary, as repulsive as certain ancient laws and punishments may now appear to us, to those who have managed to evolve spiritually, contemporary standards and ideas of justice are equally unacceptable.

Of fundamental importance is the understanding that real justice cannot be based upon a direct assessment of actions. Instead, it must be determined by *the extent to which an individual is striving to obey his conscience* – the effort or lack of effort of one to follow the dictates of one's own standard of values and not those of someone else! This all-important factor of conscience is in turn regulated by two others: an individual's *level of righteousness* – a level which is determined somewhat by birth; and the *influence of environment* upon shaping one's conscience.

But upon what principle do we presently find justice to be based? To a large degree it rests upon the idea of "fixed" penalties – *such a vital factor as conscience is completely overlooked, if not openly defied!* True justice is, of course, something that can never be rendered by man, as it is impossible for us to assess the various factors involved; but it is certainly within our ability to come much closer to this objective than we have managed in the past. Far too little effort has been made to acknowledge the role of conscience. Environment and the extent to which one has violated conscience are often sadly neglected or considered of minor importance. In general, it is deemed more expedient to adhere to the unjust but far less demanding practice of "fixed" penalties.

The Social Order

But an evaluation of sin still leaves unsolved problems. Upon what criterion should punishment be based? Exactly what is the real purpose of justice? Is it primarily to exact vengeance? If so, then it is difficult to see how we can permit such a procedure and still consider ourselves to be spiritually mature. On the other hand, if the idea is to bring one to repentance and to impart a measure of spiritual enlightenment, then why is it that many souls are kept in prison (or even executed) when they have clearly come to repentance and would be useful members of society?

Moreover, to top this atmosphere of vindictiveness we have the common practice of "setting an example." This widespread habit of imposing extra-heavy penalties for the purpose of "warning" others has no place in any enlightened civilization. It is plainly an unethical practice that only the spiritually immature can support. Constituting a personal injustice to those concerned, it is a wrong approach to the situation. Two wrongs never make a right and to resort to coercion and threats in order to reduce crime is a poor substitute for the real remedy: a higher perspective and/or a better environment. There is only one cure for a criminal, and that must be to replace an inferior outlook with one of a less selfish and more acceptable nature.

Contrary to popular opinion, a sinner is not a healthy person with a "debt" to pay to society; *he is one who is sick spiritually!* Our penal philosophy should exclude primitive ideas of specific "time" sentences, along with all vestiges of the archaic "eye-for-an-eye" mentality, and instead concentrate upon the real task of uncovering and correcting an individual's spiritual weakness. A person finds himself (or should find himself) in prison for only one reason – namely, because he is ailing in mind or spirit. Other than affording a very necessary measure of protection to its citizens, through the process of confining dangerous psychopaths and the like, our penal institutions should owe their very existence to the purpose of helping inmates to overcome mental and spiritual deficiencies.

But just as there is every justification for removing individuals from society if they are a menace to the well-being of others, *it is equally important that such persons be forbidden to return to society unless there is reasonable assurance that they have conquered their weaknesses!* To allow freedom to an unrepentant criminal is just as foolish as refusal to lock up the offender in the first place.

As if to add insult to injury, we willingly tolerate the spectacle of prisoners being confined for long periods of time and at very considerable public expense. Contributing nothing useful to society for many wasted years, they are all too often released into the world as hardened individuals with a resentful attitude and little prospect for successful assimilation and gainful employment. A more sensible policy would be to have them pay their way – and their victims – by doing useful work in State-owned industries and farms. In exchange for this labor they could learn productive skills and, in conjunction with appropriate psychiatric and spiritual counseling, eventually become an asset to society after literally earning their freedom.

As matters stand, it is really incredible that society would allow itself to be victimized twice by the same criminal. First, we have crime victims with little hope of receiving any form of compensation. Second, the general populace is saddled with the burden of paying enormous sums to maintain criminals behind bars. There is just no excuse to condone such a wasteful policy when the vast majority of inmates could be made to serve society in a useful manner. (Should there be a refusal to cooperate without legitimate cause then, quite simply, the prisoner would not eat.) Invariably, by reason of their lack of vision, it is evident that many of our highly paid legislators deserve to be removed from office on grounds of incompetence. (In cases where the death penalty has been prescribed, it would often make sense to work such a person for life, with a "poison pill" option readily available should it be so desired.)

There is yet another way in which our current method of dispensing justice leaves much to be desired. In courtrooms featuring defense lawyers pitted against prosecutors, the outcome is frequently biased in favor of wealthy defendants. To compound an unfair situation, there is generally no provision to compensate innocent parties for false charges. For a civilization that can send astronauts to the Moon, one would think that the same degree of ingenuity might be displayed with regard to such mundane issues.

Society and Religion

As an influence upon the minds and policies of man, the prevailing religious atmosphere has generally been of dubious value. Beneficial at times, it has also proven to be a source of much confusion and has caused a great deal of needless suffering. While perhaps helping some souls to obtain a more civilized outlook upon life, history has most certainly revealed many undesirable side effects. It has frequently acted to hinder the advance of science and knowledge, thereby imposing a false and inferior philosophy upon the masses. It has often led to oppression of individuals choosing to obey conscience in their quest for Truth. Indeed, it has even served to instigate many wars! In spite of its considerable influence, it cannot be said to have done much in solving many of the world's problems which are centered upon inequality of wealth.

Obviously, there has existed a warped concept as to what really constitutes true religion. The lamentable truth is that, even today, there are large numbers who are deceived in their understanding of the requirements of *true religion*, and who are oblivious to the serious shortcomings of organized religion.

True Religion is in the Heart:

In reality, true and undistorted religion is simplicity in itself. It does not embrace ceremonies or rituals of any description; nor does it necessitate one's belonging to a church or religious organization. Instead, *true religion is solely in the heart!* It consists simply of a desire to better our spiritual selves. All else is seen to be quite superfluous. Pure religion has nothing at all to do with accepting some traditional set of beliefs. Rather, there is a personal obligation to develop a genuine sense of universal love, obedience to conscience and a sincere desire to ever seek and live the Truth – regardless of the cost.

Paul, that dynamic biblical personality, expresses quite well the importance of nurturing a sense of *universal love,* in preference to other virtues like increased knowledge and a firm faith:

"If I speak in the tongues of men and of angels, but have not love, I am a noisy gong or a clanging cymbal. And if I have prophetic powers, and understand all mysteries and all knowledge, and if I have all faith, so as to remove mountains, but have not love, I am nothing. ... Love never ends; as for tongues, they will cease; as for knowledge, it will pass away. ... So faith, hope, love abide, these three, but the greatest of these is love."
(I Corinthians 13:1-2,8,13)

A more realistic interpretation of the cosmos is certainly to be treasured; but then again it must be remembered that at death one's knowledge or particular faith "will pass away" – simply because this is not actually an integral feature (but at best only an attribute) of the real self or soul. Man's earthly and imperfect views must invariably die with him; but his sense of *love* can never be affected, and will subsequently determine his future manifestation.

It may come as somewhat of a shock to some devoutly religious souls to learn that the multitude of ceremonies, rituals and dogmas presently held in esteem actually profit a man nothing; but it is nevertheless true, for they are the ideas and product of mortal man and clearly bear his mark. Indeed, all of the ceremonies, rituals and mass vocalization now fostered by our churches are essentially meaningless and to no avail. The only thing which really counts is the *sincerity* involved. It is, in fact, very often distracting and a complete waste of time to comply with the burdensome requirements of the various religious ceremonies currently in vogue.

If man would only realize that he owes no allegiance to anyone or anything except his own *conscience,* priestly influence would no longer bind him to outworn tradition, grievous to be borne, and to insidious dogma enslaving him and depriving him of his rightful heritage: a degree of freedom which Truth alone is able to provide. (Likewise, nationalism, or a foolish and puffed-up pride of one's country, would reveal itself to be exactly what it is: a vain and rather stupid sense of self-centeredness which unscrupulous leaders have been all too aware of in the past and all too eager to foster for their own evil ends.)

Just as the world once found it difficult to accept the idea of one invisible God, and clung to materialistic idol deities, so there exists today a similar tendency to adopt a "physical" approach with respect to communion with God. For instance, such an outward show of emotion as mass praising and praying – a basic mainstay of orthodox religion – is an unnecessary and wasted display of reverence. All that is needful is that we should develop the constructive habit of retiring periodically from the mad rush and excitement of the world so that our inner thoughts may be free to express themselves and where, by serious contemplation, we may thus find a new and more enlightened outlook upon life. In brief, true prayer consists simply of *deep and serious thought!* Solitude and genuine meditation are indispensable to the success of true prayer, which is fundamentally a highly *personal* undertaking. This often misunderstood principle is well enunciated in words ascribed to Jesus:

"And when you pray, you must not be like the hypocrites; for they love to stand and pray in the synagogues and at the street corners, that they may be seen by men. Truly, I say to you, they have their reward. But when you pray, go into your room and shut the door and pray to your Father who is

in secret; and your Father who sees in secret will reward you." (Matthew 6:5-6)

Another common misconception regarding prayer is the peculiar belief that one's prayers could influence God into granting favors – including intervention on behalf of others! Surprisingly, few appear to have stopped to weigh the momentous consequences of accepting such a view. Can we not see that no intervention of any sort is possible without contradicting the very laws of nature? Moreover, is it not clear that the entire concept of God intervening to assist one, at the wishes or prayers of another, is nothing short of blasphemy against the just nature of the spiritual cosmos, which demands that one be held responsible for one's own actions? Can we not see the injustice of the very notion, and understand that if the desires of one could so induce God to find favor with another it would lead to absolute chaos? Those instances in the Bible where it is advised to "pray for those who persecute you" (Matthew 5:44) were really intended to convey the following message: *Seek always to improve your level of righteousness by thinking unselfish thoughts.* (The noble act of thinking of the welfare of another is itself an indication of an increased level of righteousness and is thus an action which is to be encouraged.)

The conclusion, therefore, to be drawn from the idea of praying for others is that while it can obviously have no effect whatsoever in securing God's intervention, it can nevertheless be of definite value *to the one doing the praying!* Prayer is indeed seen to be a personal undertaking.

Society and Organized Religion:

Although criticism is something to which we all have a natural aversion, the fact remains that it is often essential if we would seek to advance the human race. In the past, organized religion has been guilty of propagating so many erroneous doctrines that one might well suspect they may have done more overall harm than good. Nor can the influence of many churches be considered an asset in this present day and age, since a host of false dogmas may be cited which pose serious obstacles to progress.

Some current examples of the public mischief wrought in the name of religion include a senseless objection to the practice of birth control, along with a totally unfounded attitude which has arisen against any form of abortion. Not only is it abundantly clear that the problem of a population explosion must soon necessitate universal acceptance of birth control as the sole alternative to utter chaos, but it is amazing that such an absurd doctrine could persist in this modern age. Invariably, the ludicrous dogma of papal infallibility has precluded all hope that any pope will be able to rectify past acts of stupidity!

Equally surprising is why anyone would tolerate the idea that a microscopic fetus is as valuable as a human life. Obviously, this theological blunder must be traced to a misunderstanding of how our souls are incarnated. For instead of becoming manifest at the moment of conception, it must now be clear that this spirit does not enter the body until some crucial – and much later stage – in the brain's development. Thus a tiny immature fetus has no greater value than any other part of our physical structure, and the traditional objection to abortion is seen to be quite baseless. (Actually, the question as to when the soul enters the body is really of little more than academic interest, since the aborted spirit of a minute fetus has not yet acquired the faculty of memory and experienced time – hence, for all essential purposes, *it has never been reincarnated!* The fact that it must promptly enter into another embryo, of like status, must serve to preserve Cosmic Justice.) It is surely far better to be born healthy to parents who truly desire a child, and who are in a position to care for it, than to risk being born either unwanted or deformed!

Another major criticism of orthodox religion centers on its negative attitude regarding the subject of evolution. The open hostility of the Roman Catholic Church toward the revelations of the now-renowned Galileo, for instance, is well-known to history students and affords a typical illustration of the blinding influence posed by church dogma. Today, science is no longer the frail child of early times; it has grown up and become a powerful factor in our lives. But even so, and despite the fact that it has revealed dogma after dogma to be nothing but a vain and foolish product of man's imagination, in general our churches have yet to overcome their reluctance to accept freely the findings of science.

The initial reaction of theology to the theory of evolution was one of bitter opposition; and even today this medieval prejudice has by no means disappeared. On the whole, the present attitude of the churches toward evolution ranges from one of unhappy (and no more than partial) recognition, to an outright rejection and a last desperate attempt to cling to an outworn and unethical redemptionist doctrine. The fact that the basic message of the Scriptures is in full agreement with the principle of evolution – which is really a spiritual process fundamentally – appears to have

The Social Order

escaped notice. Let us hope that our theologians will soon come to understand that evolution is not to be feared. On the contrary, when properly understood it will prove to be their greatest ally.

Unfortunately, any problems over evolution do not end quite so innocently as a mere difference of opinion on how life and the world came into being. Actually, the trouble is just beginning. Our churches fail to consider that their false teachings are in direct conflict with the minds of a younger generation exposed to the now-powerful influence of science. Unrealistic and obscure doctrines can only be expected to induce widespread disillusionment among those who have been forced to accept evolution as a truth. Furthermore, since theology has often branded evolution as a materialistic philosophy which denies all hope of a future life, it must follow that many of those accepting evolution *will really have been persuaded by these same churches to adopt a materialistic outlook!* It is to be trusted that theologians will soon come to realize this, and that the dynamic principle of evolution will receive due recognition.

The duties of any worthwhile religious body are unmistakable with regard to the State. It must not seek to use its influence to impose, in any way, its own particular doctrine upon the life of a nation to the exclusion of rival views – *complete freedom of religious thought must be maintained at all costs!* But neither can it afford, without betraying its moral responsibilities, to refrain from criticizing any system of government which does not comply with the principle of true democracy, and which fails to provide an environment conducive to the development of the human soul. Nor must it neglect to press for the security and freedom of all, a higher degree of justice and a fair distribution of material wealth. The social order exerts far too much influence over the life of an individual to be ignored. It is often just as essential to strive to replace a state of economic rivalry and distrust with one of mutual love and harmony, as it is to preach a more desirable personal moral philosophy.

There is, of course, one basic explanation for the many failings of organized religion – namely, *their dogmatic structure!* In short, their strict adherence to dogma prevents them from accepting a new and superior point of view if it should happen to conflict in any way with tradition. This is the one fundamental weakness of our churches, and therein is to be found the reason why they must one day submit to the advance of knowledge. For it is obvious that they cannot continue to support present dogmas much longer in the face of rapidly mounting criticism; and once they are forced to launch a serious and unbiased investigation into the mysteries of the universe it must follow that organized religion as we know it will cease to exist. Religion will have become a personal responsibility – which it has always been.

Thus, the value of an open-minded conscience is seen to be of truly fundamental importance. One may graduate from all recognized centers of higher learning, be exceptionally versed in tradition and respected highly of men and yet still possess a perspective inferior to that of one who can boast of no such qualifications, but who is nevertheless blessed with much more: *the priceless advantage of free thought and the fortitude to follow conscience – regardless of where it might lead!*

The Economic System

There are two fundamentally opposed economic ideologies upon which a social order may be based – specifically, capitalism and communism. In turn, the principle of *democracy* is of overwhelming importance to the proper functioning of either system.

Capitalism as a System:

Although certain aspects of capitalism go back in history to the first ideas of private ownership, it was only with the Industrial Revolution of comparatively recent times that this system really began to assume its present structure. Evolving, upon occasion, by the slow process of trial and error, many capitalist principles are merely a passing phase in man's quest for economic harmony. As a result, there may be witnessed considerable variation in the extent to which certain features have been retained or modified by different nations. In many instances ideas have been incorporated which might best be described as attributes of other economic systems. An example of this would be attempts to inject at least some measure of stability and security through government-sponsored pension plans and insurance against the specter of unemployment. A few of the more progressive capitalist societies have also adopted schemes of health insurance. Thus, it will be seen that the harshness of a purely capitalist economy has, to some degree, been improved by social reforms in which the interests of the public are not completely overlooked. In effect, one nation may be observed to embrace a number of socialist principles although still retaining an economy which is essentially capitalistic in nature; while another government may exhibit little change from the primitive notion of equating authority with personal wealth.

In spite of all these modifications, however, there remains many important features of this economic system that deserve criticism and which cannot be remedied until such sweeping reforms have been made as to literally convert it into a radically different social order. For reasons of ethics alone capitalism must stand condemned, as it is seen to be characterized by an unmistakable spirit of materialism and just plain selfishness. On economic grounds it must also be rejected, if for no other reason than experience has proven that it will not work in a fair and equitable manner. Moreover, left to itself, it will invariably lead to the spectacle of imperialism! (A major cause of wars since the rise of capitalism, imperialism may be defined as "a need by the wealthy to invest surplus funds in business enterprises abroad." Expecting to reap much greater profits from such investments, there is naturally a strong temptation to control the governments of the various nations involved. Historically, the First World War may be cited as a classic example of a war that was foolishly inspired by imperialistic capitalism.)

The motivating principle of capitalism is material wealth, the acquisition of which is held to be of prime importance. All actions, policies and production of goods by industry are prompted by the idea of *personal profit* – quite irrespective of whether they are justified ethically! Instead of a spirit of cooperation there will exist a situation in which one is forced to compete against his neighbor. As a result of the lack of planning which prevails, men are frequently compelled to seize whatever means of employment the system chances to provide – whether such employment be an asset to the community, a senseless waste of time and energy or even a liability to society itself. Furthermore, it strongly encourages an unfair distribution of wealth, with all the resulting evils and injustices which invariably arise when certain individuals are permitted to acquire wealth and power *at the direct expense of others who must go without!*

Nor can much be said in favor of the traditional law of supply and demand as a criterion for establishing prices, which should really be determined – if at all feasible – by such considerations as the cost of production and the ability of the end user to pay. The idea of trying to extract as much as possible from another is clearly rooted in selfishness. In addition to being morally wrong, it is also far too conducive to unfair fluctuations and extortion to serve as a basis for the ideal economic system.

The very nature of capitalism is such that it is plagued with periodic recessions and depressions, the only alternative to which has been huge peacetime military expenditures – or the specter of war, an even more deplorable state of affairs which the system also tends to foster. This is an established fact borne out by numerous instances in both past and present events, and it is of little credit to our generation that so many should condone this economic weakness, with all its serious faults, through lack of knowledge. Business cycles in which prices, wages and employment fluctuate in a most undesirable manner are not due to chance or accident, but have very definite causes arising from defects inherent in the actual system itself. For instance, is it not a strange coincidence that the Great Depression of 1929 to 1939, which was only terminated by a war, was worldwide in scope and embraced virtually all capitalist nations?

But why are capitalist societies flung periodically into a state of poverty amidst plenty? Why must people be thrown out of work, lose their homes, their health and their very perspective in life? Quite regardless of excuses which may be offered, there is just one basic underlying cause, and this may be traced to *inability of the public to purchase all the goods produced by modern industry*. Poverty is seen to exist, not due to a lack of goods, but rather to the fact that there is *insufficient purchasing power in the hands of the masses!* It is a fundamental rule of economics that the supply of money in free circulation must be consistent with the value of available consumer's goods (i.e., homes, furniture, food, clothing, automobiles, etc.). Failure to place adequate purchasing power in the hands of the common people must soon lead to a state of depression. Conversely, a state in which there is more money than goods invites inflation. In order to achieve a stable economy this balance must be maintained at all times, and any inability to do so is a clear reflection of unsound features of the prevailing system.

Under normal circumstances it is merely a question of time until a capitalist economy will find itself in such serious difficulties that, unless additional purchasing power is soon injected into the system, it must surely plunge into the depths of a depression. This chronic inability of the public to purchase all of the available consumer's goods may be traced, in essence, to an *unfair distribution of income* that sees large sums of undeserved money ending up in the pockets of a privileged class owning the means of production. Should this wealth fail to redeem a corresponding value of consumer's goods, the system will then be in danger of collapse.

While it is possible to restore some of this lost capability by way of plowing back excess profits into capital investments (i.e., new factories and additional means of production), it is also a fact that there is no way much of this money will filter down to the needy majority. There is a definite limit as to how many new manufacturing plants may be put to proper use – especially when the public is unable to purchase all the goods produced by their existing factories! (Under such conditions the threat of imperialism begins to loom on the horizon.) What is required in times of high unemployment is not tax cuts to business corporations and the rich (ostensibly to encourage capital investment), *but an increased money supply to the poor so that they might purchase a backlog of unsold goods!*

The very nature of a business slump is such that, once started, it tends to gain momentum. Once production is restricted because the public lacks sufficient purchasing power, large numbers of employees will be thrown out of work, thereby serving to add to the trouble by further reducing the ability of consumers to purchase. With high unemployment, wages and prices begin falling and hoarding of what little purchasing power there is will appear, since during a recession businessmen and consumers alike will hesitate to buy today if prices will be lower tomorrow. Banks become panicky and commence calling in loans, forcing many borrowers into bankruptcy. As a consequence, debt-burdened merchants are compelled to unload their unsold stocks of goods on the market at sacrifice prices, driving prices and wages down still further. In the face of widespread unemployment, millions will be unable to meet payments on mortgage and other loans, which still bear the high interest rates from brief periods of earlier prosperity. Needless poverty and misery abound in a land which is capable of providing a rich and enjoyable life for all, simply because the very structure of the economic system is hopelessly at fault! (It used to be that capitalist business cycles were characterized by periods of either recession or inflation. As of late, our governments seem to have found a way to give us *both* evils at the same time!)

What forces are responsible for eventually lifting a nation out of the depths of a depression? Alas, history has shown that the solution most frequently adopted is often worse than the original complaint. A state of economic prosperity usually begins, of all times, *immediately after a nation becomes involved militarily!* The reason for a war (or threat of war) working such a miraculous cure is not very difficult to find. Not only does it create an acute demand for expensive weapons of destruction, but it also absorbs multitudes of hitherto unemployed persons into the armed services. Large sums of money soon find their way into the economy of a nation because the wages paid to produce munitions – which constitute a complete waste of labor to the consumer – *represent a pure gain in the amount of cash available to purchase consumer's goods.* The trouble is not now a lack of money, but often too much of it (although most savings are still in the hands of the privileged few) and unless appropriate taxation is imposed, runaway inflation is bound to follow.

Thus, a sure cure for a depression is to resort to war. If this is not feasible, then a major armament drive, in which large numbers of young people are seduced (or forced) into armies and the public is made to bear the burden of supplying modern weaponry, will prove just as successful in saving the system from economic collapse. While clearly disgusting, such expedients will nevertheless serve to reduce unemployment and inject badly needed purchasing power. But what sort of an economic system is this which works best during a time of war, or when a fortuitous excuse can be found to rearm? Clearly, pure and unmodified capitalism leaves much to be desired.

Communism as a System:

In spite of all the negative publicity given to communism over the years, there are relatively few inhabitants of capitalist nations who can be said to possess a realistic insight into many aspects of this system. But such ignorance is perhaps to be expected in countries where control of information is largely in the hands of factions quite strongly opposed to the principle of public ownership. As a matter of fact, it is questionable whether present condemnation and hatred of communism has been fostered as much by those genuine in their sympathy for injustices under such regimes, as it has by others who have much to lose should it ever be decided that industry and wealth rightfully belong to all, and not to a privileged few.

But admitting that reports about communist nations are often "edited" in a highly biased manner by capitalist news media, the fact still remains that as a social order discontent has generally been the rule and not the exception. The idea in itself of all individuals sharing the resources of a nation constitutes a very noble philosophy, and one that can hardly be criticized on grounds of either ethics or economics. However, valid criticism arises, not as a result of public ownership, but rather because there is frequently a distinct lack of such! While it may be true that industry and commerce have been taken out of private hands and placed in control of the State, it is also a fact that in many instances the State has not been the people, but a clique of authorities terminating in a virtual dictatorship. For this very reason, it is not even certain that a true state of communism has ever existed, as this principle actually implies public ownership and control by the people themselves. (A more accurate description might, therefore, embrace such terms as "quasi-communism" or "totalitarian communism.")

It is here, in the inability of citizens to formulate their own policies, where the real objection lies. Whenever a body of people permit themselves to be dominated by a clique who will not allow any criticism of their actions, it is more or less inevitable that concern for personal feelings and even human lives will be placed in jeopardy. Instead of regarding man as a sacred spiritual entity placed here on this planet for the express purpose of evolving spiritually, such a totalitarian regime will tend to look upon one as important and an asset only to the extent that his views coincide with the aspirations of the State – the will of a virtual dictatorship – and insofar as his labor may be utilized by said State. A person will be brought up to accept authority and compelled to accept values as emphasized by a group of "leaders" who may have little sense of values. *Respect for an individual's rights and the sanctity of conscience is in danger of falling to a new low!* It remains for the future to decide whether present totalitarian communist leaderships will overcome existing objections and allow freedoms so sadly lacking in the past.

And yet, as oppressive and dictatorial as may have been certain quasi-communist regimes, one slight but vitally important change would make all the difference in the world. If only it were recognized that the State exists for the benefit of the people (and not the people for the State) most of the many inadequacies now so evident in so-called communist societies would soon vanish. In short, establishment of a genuine democracy is very likely a dire necessity before objections to current alleged communist governments can be overcome.

Unfortunately, reaction of the major capitalist nations toward the plight of such oppressed masses has been a great disappointment. Not only have they exhibited a general attitude of indifference, but there have been many instances in which they have given military support to reactionary regimes opposed to social improvement. The only sure way to halt the expansion of totalitarian communism is by establishing a chain of democracies in which wealth and power are not permitted to be concentrated so unjustly in the hands of an undeserving leadership. Failure to achieve this end may well produce serious consequences; and any infamous role played by our capitalist-based governments will be duly acknowledged by posterity!

The Democratic Principle:

While the principle of democracy must be regarded as crucial to the functioning of a just society, *it cannot be said that a genuine democracy is likely or even possible under a capitalist economy!* It may be true that the privilege to vote and to elect representatives constitutes a definite measure of democracy, but this is hardly a guarantee of an acceptable social order, nor evidence of the existence of a real democracy. On the contrary, while this principle is still in its infancy, the results so far can only be described as less than encouraging. Essentially, success of the democratic process hinges upon a combination of two factors:

1) The overall desire of a nation's inhabitants to work for a better social order; and

2) The extent to which the public is allowed to obtain an uncensored perspective.

The privilege of casting one's vote is a personal right to be cherished, posing a moral responsibility to the individual. Since the vote of an unintelligent and indifferent citizen carries just as much weight as that of one who is more enlightened, it is imperative to refrain from voting blindly.* To vote for a candidate primarily because he or she has received much favorable publicity, without ever making a sincere effort to comprehend the policies represented, is a grave mistake that will rebound against the voter and against those who are truly interested in moulding a worthwhile society. As long as we have a situation in which only a minority are earnest in their endeavor to achieve social reforms, it is inevitable that democracy should prove largely a failure. It is undoubtedly a great shame and an injustice that can only be remedied in the next world that the more spiritually evolved must suffer for the slothful and uncaring attitude of others; but such is life in this mundane phase of evolution.

The reason why capitalism is so unfavorable to the existence of a sound democracy is due, in large measure, to the serious inequality of wealth which it fosters. Inasmuch as an individual's perspective owes much to the

---

* It is surely more prudent to limit the right to vote to those capable of passing a stiff test designed to weed out the incompetent and the misinformed. Ability to discern *principles* is of the utmost importance in any election. Most assuredly, there is little merit in basing intelligence solely upon some arbitrary age. The privilege of voting is a right that should be earned!

influence of environment, it is to be expected that such media as newspapers, magazines, radio and television will constitute a most important source of persuasion. But whose views do these influential organs reflect? Almost without exception they express the *personal views and aspirations of their owners,* who are invariably limited to wealthy individuals who usually manage to see eye-to-eye with big business interests – for the simple reason that they *are* big business! If it should suit their purpose they are able to tactfully withhold or even color news items that would tend to cast an unfavorable light upon their ambitions. On the other hand, they are able to magnify, often out of all proportion, certain trivial occurrences that happen to coincide with their policies. In short, by controlling the main avenues of information it is frequently possible to restrain those passions it desires to suppress, and to incite those passions that are likely to prove advantageous. (Ironically, before the advent of mass media, and the necessity of purchasing much flattering publicity, the democratic process stood a better chance of working.)

As regrettable as this one-sided control of information may be, the trouble extends to the actual choice of political representatives. Here money also plays a predominant role. It so happens that one lacking financially has little hope of success without receiving publicity and monetary assistance from some exterior source. As a result, it is not in the least surprising that most of our politicians and political parties owe allegiance to business interests, whose selfish motives have led them to provide such contributions. *But nobody is likely to give something for nothing!* It is, in fact, no accident that genuine democracy is little more than an illusion and that, on the whole, our governments have failed to properly reflect the wishes of its citizens. It may well be considered a disaster of our times that so many freedom-loving nations hold periodic elections in which the public has little or no opportunity to elect a truly acceptable government – simply because there is not a sufficient number of worthy candidates from which to choose!

This point is amply demonstrated upon examination of the supposed "rivalry" between political groups. With very few welcome exceptions, elections in most capitalist nations feature two political parties divided by differences which are frequently far more imaginary than real. In all such instances they are financed by big business firms who clearly expect a return on their investments, which is why many will contribute to *both* parties! (No matter who is elected they stand to receive favors.) It is hardly a coincidence that virtually all elected leaders are wealthy individuals, or puppets of very wealthy interests. This is also the type of person most likely to possess the undesirable trait of selfishness!

Nevertheless, in spite of the many obvious economic, social and political defects of capitalism, the public itself is not without a definite measure of responsibility. As long as a nation has the freedom to vote, regardless of any unfair influence imposed by wealth, it is within its power to overcome these obstacles and to create a society of its own choosing. Social improvements can be obtained under such circumstances, but only at the cost of much effort and through the acquisition of an enriched perspective on the part of the masses. If we truly desire a better social order we must not be afraid to work and sacrifice to achieve this noble end.

In Search of a World Society

While it is to be concluded that neither quasi-communism nor capitalism offer any real hope of combining economic prosperity with freedom and justice, upon what basis may world harmony be founded? Moreover, what principles of ethics must be evoked to assure a society that is acceptable morally; and what changes of an economic nature must be adopted to provide stability and security?

A New United Nations:

Almost since its inception, it has been apparent that the United Nations, as currently formulated, is far too weak and ineffectual to guarantee world peace. In many respects a carbon copy of its failed predecessor, the League of Nations, this body has proven to be no more successful in settling international disputes and in achieving world disarmament. A multitude of wars and atrocities have been allowed to ravage the globe without interference, in open defiance of its very presence. The momentous question now confronting us is whether there is any way it may be revitalized and given the power to deal with aggression and injustice.

By far the most pressing issue facing the world today involves the abolition of nuclear weaponry and germ/chemical warfare. Taken either separately or together, these twin evils threaten the annihilation of Earth society. It is a highly alarming state of affairs to realize that a multitude of minor nations will possess one or more of these fearful capabilities either before or shortly after the dawn of the 21st century! What makes the situation

The Social Order

even more ominous is that history has repeatedly demonstrated man's willingness to resort to acts of sheer terrorism, sometimes with little provocation or any forethought as to consequences. It requires no great taxing of the imagination to envisage the wholesale carnage once some mentally unbalanced leader (and there always seems to be a number of such characters in power) begins pushing buttons. Following an inevitable retaliation, the stage would be set for uncontrolled escalation. The time to establish a *World Government*, which would positively – and effectively – outlaw weapons of this nature, *is now, before it becomes too late! To delay is to risk global suicide.*

This foolish reluctance to adopt a firm disarmament policy, by our present inept United Nations, is most distressing. Can it not be seen that further procrastination will soon result in a rapidly growing list of small nations (many with dictatorial leaders) being able to deliver a staggering amount of mass destruction? To compound the problem, once a minor nation acquires the Bomb (which can be delivered very cheaply and effectively by means of a "suitcase") an unscrupulous dictator will have little incentive to give it up and rely upon mere conventional arms – a field in which it cannot compete! Undoubtedly, gross stupidity resides in high places.

There is really only one solution to the potential threat now before us. A *World Government,* possibly within the framework of a new and much-reformed United Nations, is essential to the establishment of a *World Army*. Comprised of carefully screened volunteers from many nations, and pledged to fight aggression and injustice wherever it is seen to exist, it would initially retain the only legal stockpile of atomic warheads. Upon successfully decommissioning the last nuclear missile submarine and having taken every measure to neutralize the possibility of germ/chemical warfare, it would then press for a ban on conventional arms – on the grounds that this one World Army would act swiftly and decisively to crush any form of aggression. Henceforth, no nation on Earth would have reason or need to squander valuable resources and manpower on behalf of a senseless military capacity.

With the overwhelming superiority of such a World Army, the United Nations would, for the first time in its history, be in a position to do something about the many injustices posed by our planet's assorted petty dictatorships. In fact, a prime responsibility and priority of this hitherto passive world organization must entail a campaign to overthrow all governments which fail to give the democratic process a fair chance. To this end, a powerful World Army – acting as a truly International Police Force – is clearly essential to global security.

Based upon proportional representation among nations embracing the democratic principle, with no nation possessing the power of veto, the foundation would be laid for a World Government to devise a prompt and equitable solution to many of Earth's problems. The time has come for mankind to put aside silly grievances and prejudices, which lie far beneath the demands and responsibilities of the 20*th* century. What is needed is a spirit of cooperation and mutual trust, in which individuals of all creeds and nationalities are willing to grow up and *think unselfishly in terms of the entire human race!* Unification – not separation – must guide our every thought and action.

This last point cannot help but raise a rather pertinent question. Why is it that the United Nations, supposedly devoted to the task of uniting nations, has failed to press for a universal language? If ever there was a factor designed to prevent humanity from achieving harmony, it is surely the lack of a common speech. But why continue this foolish babbling, in multiple tongues, when there is absolutely no reason to tolerate such a ridiculous scenario? It stands as little credit to our witless politicians that they have not seen fit to insist upon a *world language!* Derived from the best features of many cultures, a scientific language could easily be devised and taught in the schools of every nation. Within a generation or two it could become the official tongue of the planet Earth, supplanting all others and providing a new basis for world understanding. It would also signify that, at long last, we have reached the stage where we might be justified in thinking of ourselves as prospective members of a Galactic Society – rather than a disorganized rabble inhabiting some primitive and unevolved world.

Regrettably, there is yet another new and rarely discussed global crisis looming large on the horizon – one which threatens mankind from *within* unless appropriate action is soon taken upon an international scale. Thanks to the efforts of medical science, the lives of many individuals, plagued with a wide variety of serious genetic diseases (i.e., diabetes, hemophilia, cystic fibrosis, sickle cell anemia, etc.), have been greatly extended. While this is certainly to be desired, it has nevertheless interfered with nature's evolutionary process by allowing many carriers of defective genes to live past the age of puberty – thereby passing on these defects to all succeeding generations. The end result of this intervention is bound to be *cumulative* in effect and it will not be much longer than a century or two until almost every person on this planet will carry a significant genetic defect! (What is presently needed is not so much a marriage license as a birth license, before the situation gets totally out of hand.) In conjunction with a worldwide policy of birth control, this weeding out of serious genetic defects is a problem which cannot be avoided indefinitely and must be addressed effectively – preferably upon the level of a New United Nations – in the very near future, and before quantity is allowed to completely eclipse quality!

Economic Sharing:

Contrary to popular opinion, the natural resources of our planet rightfully belong to everyone – not to those whom some arbitrary national boundary would seem to bestow a privilege. If ever we are to rise above the mentality of ignorant tribesmen, it is essential to realize that no nation has a legitimate and exclusive claim to that which has been supplied by nature. The selfishness of a few to the detriment of many is an attitude that can no longer be condoned by a civilized humanity. It is just as much an outrage for a group of oil-rich sheiks to resort to acts of extortion, as it is for other nations to refuse food to a starving populace. Similarly, the timber, fisheries and mineral resources of the world are a heritage given to *all*. The ability of nations to share their blessings is a must if there is to be a peaceful resolution of global tensions.

The task of economic reform is one that invariably raises the issue of public ownership. To what extent is it desirable to replace private enterprise with a system of democratic State ownership or control? In spite of the numerous injustices which capitalism has spawned, it must not be thought that the system is entirely without merit. On the contrary, inequalities arose chiefly because there has been no effective limitation upon the wealth and power accumulated by a small minority of citizens; and as long as a more just distribution of wealth can be assured there are many features that are beneficial in our current stage of social evolution. The right of one to own a home, farm or business establishment of reasonable size should be recognized by the State.

Nevertheless, it is also clear that a considerable amount of public ownership is essential if present objections are to be overcome and resources utilized for the benefit of all. In particular, there must be a limitation placed upon personal wealth, as it is this great inequality that leads to so many undesirable consequences. But what specific business enterprises should be owned by the public, and what actual restrictions should be placed upon private wealth and business corporations?

Limitation of wealth may be achieved by placing an initial ceiling upon the total assets that may be held by an individual. The exact figure of this proposed limitation is something which must naturally be based upon circumstances and the wealth of a nation. However, a tentative value of perhaps thirty times the yearly income of the average worker is not entirely out of line with the need to encourage private initiative on the one hand, and the necessity to achieve a more acceptable distribution of wealth on the other. Future injustice could be negated through imposition of higher succession duties, and by an income tax so designed as to confiscate entirely all income above a certain sum – possibly, an amount equal to no more than ten times the average annual wage. (The structure of such tax laws would, of course, be founded on the *progressive tax principle:* essentially, the greater the income the higher the percentage of tax – with a basic exemption sufficient to exclude those of low income.)

Public control over a portion of a nation's economy will serve, in many ways, to regulate that portion under private ownership, since any private business will be competing with the State for available labor. Thus wages and working conditions must be competitive with businesses under public domain. Except for whatever limitation may be imposed by private inability to finance a really large business, a free economy will prevail and inspire the greatest possible efficiency. If public control of a particular industry should prove more efficient and more desirable than private control, the former will eventually replace the latter on grounds of merit. Conversely, if private initiative can be shown to be equal or superior in a specific business, then there is no reason to change to State ownership.

By far the most important business demanding public ownership is that of *finance – a medium which totally regulates the very economy of a nation!* It is incredible that anything so vital should be entrusted to private ownership. Future generations will surely be amazed by policies which we presently condone. Not only does private control of finance often result in a poor channeling of funds to worthy projects, but the established practice of paying interest for the use of money constitutes an unnecessary hardship for many, and serves to lower the standard of living for all but the wealthy. The idea of charging interest for borrowed money has little justification, either ethically or from the point of view of economic stability. Morally speaking, any money obtained by way of interest cannot be considered earned, as no genuine contribution to society will have been made. On economic grounds there is certainly no conflict with the requirement that the money supply must always balance available consumer's goods, since all loans must be repaid. The concept of demanding interest in return for the use of money constitutes one of the main principles upon which the capitalist system is based. In effect, it is actually little more than a device to place an unnecessary burden upon the shoulders of the public. The only ones who really stand to gain are those who possess wealth!

Admitting, therefore, that the principle of usury is unjust and undesirable, why is it condoned by society? Incredibly, there would appear to exist a widespread belief that there is actually a necessity to borrow money from private sources. To illustrate this ridiculous obsession we have only to observe the various levels of our government – *for even they have somehow managed to fall into the greedy clutches of private financiers!* Instead of simply printing and loaning money (which must be repaid) free of interest to itself for essential and worthwhile purposes, we witness the absurd spectacle of governments first permitting such funds to pass through the hands of private financial institutions, who naturally seize upon this opportunity to ask interest. Thus there exists a situation whereby sanction from government is necessary to print money, but in order for the various levels of government to obtain these funds it must consent to pay "middlemen" large sums in the form of unnecessary interest to give it back to itself! How crazy can we get?

The abolition or restriction of usury would result in enormous benefits to municipalities and private citizens alike. As a direct consequence of State control of finance, money for constructive and desirable purposes would be available for only minor administrative fees. Loans for hospitals, schools, municipal buildings, public works, roads, etc., could be obtained almost free of charge, thereby greatly reducing taxes and prices. But of truly immense benefit to the average individual would be the priceless advantage of being able to secure virtually interest-free loans for the erection or purchase of homes and farms.

Yet another injustice of the monetary policies of capitalism is the practice of deficit financing. It is a fact of sound economics that stability requires a balanced national budget. Quite simply, there is no excuse to let the money supply get out of synchronization with available consumers' goods. In times of war or rearmament, when surplus money is injected into the system, the usual procedure is to avoid recovering the full amount of deserved tax – thereby producing inflation which is alleviated momentarily through the supposedly patriotic gesture of issuing government savings bonds. This is surely a most despicable ploy that should be outlawed by constitution! *It is nothing less than a crooked device to allow the wealthy to escape their due responsibilities,* since it is obvious that the poor cannot be asked to pay substantially more than they already have without risking strong reaction. By this blatant subterfuge, the rich are permitted to retain excess profits and a "debt" is incurred for which there is absolutely no justification. (From an economic standpoint, it is clearly impossible for a nation to have a legitimate debt for something *already produced with its own resources!*)

The mentality of Earth politics would never cease to amaze a visitor from another planet. While many capitalist nations finally got around to initiating a variety of old age pension and unemployment insurance schemes, in order to compensate for loss of income, for some inexplicable reason they have all failed to see the merit of *adjusting retirement age* as a means of abolishing unemployment altogether! By merely lowering the age at which one could retire on pension, in times of unemployment, it would free sufficient jobs as to totally eliminate the problem. (A person would, of course, have the choice of either pension or work – but never both.) Why endure the common spectacle of disgruntled youth, torn between idleness and crime, while many older citizens – desirous of retiring, if only they could afford to do so – continue to labor and pay unnecessary taxes? In effect, older and less healthy individuals are compelled to support those more capable and more desirous of membership in the labor force! Nor can this stupid oversight be excused on grounds of cost, since the additional amount required for pensions need never exceed what is currently being spent on such props as unemployment benefits, welfare payments, and a host of essentially useless retraining and "make work" programs – of which enticing (or forcing) youth into the military may be cited as a modern approach to the problem. (By a conservative estimate, and with an average work week of less than 40 hours, it should be possible to lower the retirement age to no more than 50 years.)

A planned economy, devoid of any unfair distribution of wealth, wasteful duplication, unemployment and (above all) senseless military expenditures, would at last permit genuine peace and a great increase in world living standards. But nothing of real value comes easy. If the imperfect environment of one era is ever to be transformed into a more just and harmonious society of the future, it is bound to come only as a result of much effort on our behalf.

It is essential that we rise above the tribal level and think in terms of acquiring a *Cosmic Perspective*. Only then will it be possible to realize the glorious destiny which awaits mankind.

Index
(Parts One and Two)

INDEX

Aaron ... *155 - 156*
abortion ... *249 - 250*
Abraham ... *156, 164, 172 - 174, 176 - 177, 201*
Acts ... *159, 206, 227 - 229, 232 - 234*
Adam and Eve ... *14, 166 - 167, 169 - 170*
Ahaziah, King ... *194*
amphibians ... *5 - 7*
Ananas ... *231, 233*
Andrew ... *201*
angular momentum ... *32*
antiproton ... *30, 74*
apostles ... *165, 201*
Arp, Halton C. ... *95 - 135*
asteroids ... *31*
atoms ... *16, 18 -19, 24, 28 - 29*
Bacall, John ... *146*
Baigent, Michael ... *223*
baptism ... *200 - 201, 207*
Barabbas ... *208*
Biblical symbolism ... *154, 163 - 166*
birds ... *8 - 9*
birth control ... *249 - 250, 263*
black holes ... *37 - 38, 40, 42, 45 - 46, 75, 78 - 79, 93, 113, 117, 121, 129 - 134, 145*
blueshifts ... *110, 112*
Burbidge, Geoffrey ...*95*
Cambrian period ... *4, 5 (Fig. 1)*
capitalism ... *252 - 258, 260, 263, 265 - 266*
carbon-14 ... *13, 224 - 225, 238*
Carboniferous period ... *5 (Fig. 1), 7*
cells ... *15 - 16, 19 - 21*
Cherry, David ... *95*
"Christ" ... *157, 159, 162, 164, 195 - 199, 203 - 208, 210 - 213, 215, 218 - 219*
chromosomes ... *16 - 17*
comets ... *32*

communism ... *256 - 257, 260*
conscience ... *168 - 170, 180 - 181, 184, 188 - 189, 191, 193, 199, 206, 211, 216, 218 - 219, 240 - 241, 243 - 244, 247 - 248, 252, 257*
consumer's goods ... *254, 256, 265*
Cooke, W.J. ... *98*
creation story ... *166 - 168*
creation models:
    big bang ... *47 - 50, 71, 81 - 82, 84 - 85, 90, 101, 107, 110, 114 - 117, 135 - 136, 143 - 148*
    inflationary big bang ... *48*
    oscillating big bang ... *48*
    "special creation" ... *15*
    Steady-state (or continuous creation) ... *49 - 50, 70 - 71, 77, 81, 83, 85, 93, 124, 127, 143 -144, 148*
Cretaceous period ... *5 (Fig. 1), 9*
D/R factor ... *61, 62 (Fig. 3), 65, 94 - 95, 96 (Fig. 11), 97 - 99, 104 - 105, 110, 112, 124, 135 - 136, 145 - 146, 148*
D/R quanta ... *95, 97 - 99, 103, 109, 111 - 114, 116, 148*
Damascus ... *164, 206, 230, 233*
Daniel ... *191 - 193, 212 - 213*
David ... *194*
Dead Sea Scrolls:
    Damascus Document ... *226, 229*
    dating of ... *224 - 225*
    discovery of ... *220 - 223*
    Community Rule ... *222, 227 - 228*
    Copper Scroll ... *222, 226*
    Habakkuk Commentary ... *222, 226, 230 - 231*
    Temple Scroll ... *228 - 229*
    War Scroll ... *222, 227*
deficit financing ... *265 - 266*
democracy ... *251, 258 - 259*

Deuteronomy ... *155, 183*
Devil ... *(see Satan)*
Devonian period ... *5 (Fig. 1), 6*
dinosaurs ... *7 - 9, 21*
discrimination ... *241 - 242*
distribution of resources ... *241, 251, 263 - 264*
DNA ... *16*
Doppler shift ... *42 - 43, 115 - 116*
Earth ... *3 - 4, 9, 19*
Einstein, Albert ... *45, 49, 60*
Eisenman, Robert ... *230*
elections ... *258 - 260*
electricity ... *28 - 30, 71 - 73, 122*
electrons ... *28 - 29, 46, 57, 72 - 74, 76, 114, 122, 136*
Elijah ... *194*
Eocene epoch ... *10*
Esau ... *174 -177*
Essenes ... *221*
Esther ... *187 - 189*
evolution:
    mechanism ... *15 - 21*
    philosophical implications ... *14, 21 - 26, 153, 250 - 251*
Exodus ... *155 - 156, 179 - 184*
Ezra ... *156*
fishes ... *5 (Fig. 1), 6*
fossils ... *4, 6 - 13, 15*
free will ... *20 - 22, 169, 190, 207, 235*
Friedman, Elliott ... *157*
fusion, cosmic ... *25 - 26, 51 - 52, 57, 118 - 124, 126, 128, 130 - 134, 136*
galaxies:
    bubble-like distribution ... *63, 81 - 82, 99*
    classification ... *41*
    Local Group ... *41 - 42, 43, 98, 116, 124*

Local Supercluster ... *42, 46, 81, 107, 108 (Table 1), 109 (Table 2), 110, 112, 147 - 148*
Magellanic Clouds ... *40 - 41*
spiral structure ... *61*
superclusters ... *42, 44, 73, 75, 80 - 81*
AM 059-4024 ... *103*
AM 0328-222 ... *103, 141*
AM 2006-295 ... *102, 104, 141*
AM 2054-221 ... *104, 141*
M31 ... *41, 98*
M51 ... *42*
M63 ... *42*
M81 ... *41, 98*
M82 ... *41, 97*
M87 ... *42, 79, 109 - 110, 133, 148*
M94 ... *42*
M101 ... *42*
NGC 1232 ... *103, 141*
NGC 2403 ... *41*
NGC 2997 ... *(Jacket Cover)*
NGC 4258 ... *42*
NGC 4319 ... *97*
NGC 7331 ... *105 - 106*
NGC 7603 ... *102, 141*
Seyfert's Sextet ... *105*
Stephan's Quintet ... *105 - 106, 141*
VV 172 ... *105*
gamma rays ... *117, 250*
genes ... *16 - 21, 24*
Genesis ... *155, 165, 170 - 178*
genetic defects ... *262 - 263*
globular clusters ... *40 - 41, 110, 133, 147*
Gospels:
    John ... *158 - 159, 161, 196 - 198, 202 - 204, 208 - 211, 219*

INDEX

Luke ... *158 - 159, 165, 197, 199 - 201, 205, 207, 209, 219, 232*
Mark ... *158 - 159, 208, 219*
Matthew ... *158 - 159, 161, 164, 198 - 206, 208, 210, 212 - 214, 219, 228, 232, 248 - 149*
gravitation:
   attraction ... *51 - 56*
   nature of ... *51 - 52*
   repulsion ... *56, 58 - 61, 62 (Fig. 3), 63, 135*
   speed of ... *52 - 53, 56, 64, 113 - 115, 135*
Great Attractor ... *98*
Great Wall ... *97 - 98*
guidance ... *19 - 21*
helium ... *29, 31, 36 - 37, 76*
Hezekiah, King ... *156 - 157*
Hubble's constant (or Hubble Law) ... *43, 46, 60, 67 - 68, 78, 80, 110, 112, 124, 126 - 127, 147 - 149*
human ancestry ... *11 - 15*
human restrictions ... *240, 242 - 246*
human rights ... *240 - 245*
hydrogen ... *29, 31, 36 - 37, 47, 71, 76, 100, 111, 123, 128, 143 - 144*
imperialism ... *253*
insects ... *5 (Fig. 1), 6*
International Team ... *235 - 238*
Isaac ... *174 - 176*
isotopes ... *3, 224 - 225*
Jacob ... *156, 165, 172, 174 - 178*
James ... *162, 201, 231 - 235*
Jeremiah ... *155*
Jeroboam, King ... *156*
Jesus:
   birth scenario ... *198 - 200*
   crucifixion ... *206 - 209*
   early years ... *200 - 201*
   "miracles" ... *201 - 206*
   resurrection ... *209 - 211*

Job ... *189 - 191*
John the Baptist ... *226 - 232*
Jonah ... *164, 193, 210*
Joseph ... *176 - 178*
Joshua ... *156, 165, 184 - 185*
Josiah, King ... *155*
Judas ... *166, 207*
Judges ... *156, 185 - 186*
Jupiter ... *31*
Jurassic period ... *5 (Fig. 1), 7 - 8*
justice ... *243 - 246*
Kings (I & II) ... *156, 194*
laws:
   addition of speeds ... *45, 66 - 67, 70, 77*
   conservation of energy ... *77 - 78*
   D/R law ... *61, 62 (Fig. 3)*
   inverse-square law ... *56, 100, 135*
   Newton's third law ... *52, 73*
Lazarus ... *164, 204*
Leigh, Richard ... *223*
Lerner, Eric J. ... *116*
Leviticus ... *156*
Liar ... *226, 230, 234*
Mack, Burton L. ... *219*
mammals ... *5 (Fig. 1), 7, 9 - 11*
Marmet, Paul ... *101, 116*
Mars ... *31*
Masada ... *227*
Mercury ... *31, 60 - 61*
mesons ... *76*
Michelson-Morley experiment ... *45, 66 - 67*
microwave background ... *49, 71, 114 - 117*
military service ... *240, 256, 266*
Milky Way galaxy ... *38 - 42, 79, 85, 97 - 98, 102, 104, 106, 109 - 113, 116, 141, 145*
Miocene epoch ... *5 (Fig. 1), 11*
"missing mass" ... *60, 136, 146*
Mithraism ... *161*

273

Mohammed ... *162*
Moses ... *155 - 157, 165, 178 - 184, 200, 228, 234 - 235*
mutations ... *17 - 18, 21*
nationalism ... *239, 248*
Neptune ... *31, 60, 63*
neutrinos ... *76 - 77, 114*
"neutral quanta" ... *55 - 56, 64 - 65*
neutrons ... *29, 37, 74, 76*
New Testament:
  authorship ... *157 - 160*
  text ... *195 - 219*
  "Q" document ... *159*
New United Nations ... *240, 260 - 263*
Numbers ... *155 - 156*
Old Testament:
  authorship ... *155 - 157*
  text ... *167 - 194*
Oligocene epoch ... *5 (Fig. 1), 10*
Ordovician period ... *5 (Fig. 1), 6*
original sin ... *161, 169 - 170, 181*
Paul ... *158, 161 - 162, 164, 206, 230, 233 - 235, 247*
Pentateuch ... *155 - 157*
Peter ... *158, 166, 201, 206 - 207, 232*
Peter (I & II) ... *158 - 159*
Pilate, Pontius ... *158*
planetary systems ... *30 - 33, 128, 130 - 131*
Pliocene epoch ... *5 (Fig. 1), 11*
popes ... *162, 250*
positrons ... *30, 46, 73 - 74, 76, 114*
prayer ... *248 - 249*
primates ... *5 (Fig. 1), 11*
protons ... *29 - 30, 37, 57, 72 - 76, 122 - 124, 126, 128, 130, 133*
public ownership ... *263 - 265*
pyramid structure:
  of cosmos ... *26, 118 - 134*
  of life ... *20 - 21, 23, 25*
quarks ... *76*

quasars ... *45 - 48, 75, 79 - 81, 82 (Fig. 6), 83 - 85, 86 - 87 (Fig. 7), 88 - 89 (Fig. 8), 90, 91 (Fig. 9), 93 - 95, 96 (Fig. 11), 97, 116 - 117, 136, 143 - 144*
Qumran ... *221 - 223, 225 - 235*
radiation:
  curvature of ... *70, 85, 94 - 95, 96 (Fig. 11), 97 - 98*
  entrapment of ... *77 - 78, 85, 90*
  escape velocity of ... *66*
  nature of ... *63*
  propagation of ... *50, 63 - 67, 68 (Fig. 4), 69 (Fig. 5), 70 - 71, 77, 93 - 94, 99 - 101*
radio sources ... *49, 116*
redemption doctrine ... *161, 235, 250*
redshifts ... *43 - 46, 67, 68 (Fig. 4), 70 - 71, 81, 82 (Fig. 6), 85, 90 - 91, 92 (Fig. 10), 93 - 95, 99 - 101, 143, 147 - 148*
redshifts, discordant ... *101 - 107, 108 (Table 1), 109 - 114, 135*
Rehoboam, King ... *156*
reincarnation ... *22 - 24, 125, 128 - 129, 136, 153, 161*
relativity ... *(see Einstein)*
religion (organized) ... *163, 247 - 252*
reptiles ... *5 (Fig. 1), 7 - 9*
retirement age ... *266*
Revelation ... *159, 165, 214 - 219*
Roman Catholic Church ... *162, 250*
Sagittarius A ... *40*
Samson ... *185 - 187*
Samuel (I & II) ... *156, 194*
Satan (or Devil) ... *195 - 196, 201*
Saturn ... *31*
Silurian period ... *5 (Fig. 1), 6*
small bang ... *75 - 76, 79 - 81, 83, 144, 148*
socialism ... *252, 263 - 264*
Solomon, King ... *156 - 157*

spirit ... *16 - 21, 30, 119 - 120, 123, 125 - 128, 130 - 132, 133 (Fig. 12)*
stars:
   degenerate ... *37 - 38, 120*
   distances ... *33 - 34*
   evolution of ... *35 - 38*
   luminosities ... *34 - 35*
   main-sequence ... *34 - 35*
   masses ... *34, 38*
   neutron stars ... *34, 37 - 38, 121, 129, 145*
   pulsars ... *38*
   pulsating ... *34 - 36*
   red giants ... *34 - 36*
   sizes ... *34*
   supernovae ... *37 - 38, 75*
   white dwarf ... *34 - 37*
   Sun ... *3, 30 - 36, 63, 133*
Teacher of Righteousness ... *229 - 231, 233*
Temple, Jerusalem ... *155, 212 229, 231 - 133*
Tertiary period ... *5 (Fig. 1), 9*
Thomas ... *211*
Tifft, W.G. ... *98, 112*
"time constant" ... *125 - 129, 136*
Triassic period ... *5 (Fig. 1), 7*
Tully, R. Brent ... *107*
universal language ... *262*
universe:
   center of ... *70, 80, 84 - 85, 90, 93 - 94, 97, 100, 105, 107, 108 (Table 1), 111 - 112, 116 - 117*
   edge of ... *70, 77 - 78, 81, 84 - 85, 90 - 91, 92 (Fig. 10), 93 - 94, 97, 100 - 102, 104, 106 - 107, 108 (Table 1), 111 - 117*
   expansion of ... *42 - 45, 69 (Fig. 5), 73, 80, 126, 135, 145*
   other universes ... *118, 132*
   size of ... *80, 93, 148*
uranium ... *3*

Uranus ... *31, 60*
usury ... *264, 265*
Venus ... *31*
virial theorem ... *59*
Wicked Priest ... *230 - 231, 233*
Zealots ... *221, 225, 227, 232 - 233*

Additional Titles
by Sunstar Publishing Ltd.

- *The Name Book* by Pierre Le Rouzic
  ISBN 0-9638502-1-0 $15.95

  Numerology/Philosophy. International bestseller. Over 9,000 names with stunningly accurate descriptions of character and personality. How the sound of your name effects who you grow up to be.

- *Every Day A Miracle Happens* by Rodney Charles
  ISBN 0-9638502-0-2 $17.95

  Religious bestseller. 365 stories of miracles, both modern and historic, each associated with a day of the year. Universal calendar. Western religion.

- *Of War & Weddings* by Jerry Yellin
  ISBN 0-9638502-5-3 $17.95

  History/Religion. A moving and compelling autobiography of bitter wartime enemies who found peace through their children's marriage. Japanese history and religion.

- *Your Star Child* by Mary Mayhew
  ISBN 0-9638502-2-9 $16.95

  East/West philosophy. Combines Eastern philosophy with the birthing techniques of modern medicine, from preconception to parenting young adults.

- *Lighter Than Air* by Rodney Charles and Anna Jordan
  ISBN 0-9638502-7-X $14.95

  East/West philosophy. Historic accounts of saints, sages and holy people who possessed the ability of unaided human flight.

- *Bringing Home the Sushi* by Mark Meers
  ISBN 1-887472-05-3 $21.95

  Japanese philosophy and culture. Adventurous account of of an American businessman and his family living in '90s Japan.

- *Miracle of Names* by Clayne Conings
  ISBN 1-887472-03-7 $13.95

  Numerology and Eastern philosophy. Educational and enlightening – discover the hidden meanings and potential of names through numerology.

- *Voice for the Planet* by Anna Maria Gallo
  ISBN 1-887472-00-2 $10.95

  Religion/Ecology. This book explores the ecological practicality of native American practices.

- *Making $$$ At Home* by Darla Sims
  ISBN 1-887472-02-9 $25.00

Reference. Labor-saving directory that guides you through the process of making contacts to create a business at home.

- *Gabriel & the Remarkable Pebbles* by Carol Hovin
  ISBN 1-887472-06-1 $12.95

Children/Ecology. A lighthearted, easy-to-read fable that educates children in understanding ecological balances.

- *Searching for Camelot* by Edith Thomas
  ISBN 1-887472-08-8 $12.95

East/West philosophy. Short easy-to-read, autobiographical adventure full of inspirational life lessons.

- *The Revelations of Ho* by Dr. James Weldon
  ISBN 1-887472-09-6 $17.95

Eastern philosophy. A vivid and detailed account of the path of a modern-day seeker of enlightenment.

- *The Formula* by Dr. Vernon Sylvest
  ISBN 1-887472-10-X $21.95

Eastern philosophy/Medical research. This book demystifies the gap between medicine and mysticism, offering a ground breaking perspective on health as seen through the eyes of an eminent pathologist.

- *Jewel of the Lotus* by Bodhi Avinasha
  ISBN 1-887472-11-8 $15.95

Eastern philosophy. Tantric Path to higher consciousness. Learn to increase your energy level, heal and rejuvenate yourself through devotional relationships.

- *Elementary, My Dear* by Tree Stevens
  ISBN 1-887472-12-6 $17.95

Cooking/Health. Step-by-step, health-conscious cookbook for the beginner. Includes hundreds of time-saving menus.

- *Directory of New Age & Alternative Publications* by Darla Sims
  ISBN 1-887472-18-5 $23.95

Reference. Comprehensive listing of publications, events, organizations arranged alphabetically, by category and by location.

- *Educating Your Star Child* by Ed & Mary Mayhew
  ISBN 1-887472-17-7 $16.95

East/West philosophy. How to parent children to be smarter, wiser and happier, using internationally acclaimed mind-body intelligence techniques.

Sunstar Publishing Ltd.

- *How to be Totally Unhappy in a Peaceful World* by Gil Friedman
  ISBN 1-887472-13-4 $11.95

  Humor/Self-help. Everything you ever wanted to know about being unhappy: A complete manual with rules, exercises, a midterm and final exam. (Paper.)

- *No Justice* by Chris Raymondo
  ISBN 1-887472-14-2 $23.95

  Adventure. Based on a true story, this adventure novel provides behind the scenes insight into CIA and drug cartel operations. One of the best suspense novels of the '90s. (Cloth.)

- *The Symbolic Message of Illness* by Dr. Calin Pop
  ISBN 1-887472-16-9 $21.95

  East/West Medicine. Dr. Pop illuminates an astonishingly accurate diagnosis of our ailments and physical disorders based solely on the observation of daily habits.

- *On Wings of Light* by Ronna Herman
  ISBN 1-887472-19-3 $19.95

  New Age. Ronna Herman documents the profoundly moving and inspirational messages for her beloved Archangel Michael.

- *The Global Oracle* by Edward Tarabilda & Doug Grimes
  ISBN 1-887472-22-3 $17.95

  East/West Philosophy. A guide to the study of archetypes, with an excellent introduction to holistic living. Use this remarkable oracle for meditation, play or an aid in decision making.

- *Destiny* by Sylvia Clute
  ISBN 1-887472-21-5 $21.95

  East/West philosophy. A brilliant metaphysical mystery novel (with the ghost of George Washington) based on *A Course In Miracles*.

- *The Husband's Manual* by A. & T. Murphy
  ISBN 0-9632336-4-5 $9.00

  Self-help/Men's Issues. At last! Instructions for men on what to do and when to do it. The Husband's Manual can help a man create a satisfying, successful marriage – one he can take pride in, not just be resigned to.

- *Twin Galaxies Pinball Book of World Records* by Walter Day
  ISBN 1-887472-25-8 $12.95

  Reference. The official reference book for all Video Game and Pinball Players – this book coordinates an international schedule of tournaments that players can compete in to gain entrance into this record book.

COSMIC PERSPECTIVE

- *How to Have a Meaningful Relationship with Your Computer*
  by Sandy Berger
  ISBN 1-887472-36-3 $18.95

Computer/Self-help. A simple yet amusing guide to buying and using a computer, for beginners as well as those who need a little more encouragement.

- *The Face on Mars* by Harold W.G. Allen
  ISBN 1-887472-27-4 $12.95

Science/Fiction. A metaphysical/scientific novel based on man's first expedition to investigate the mysterious "Face" revealed by NASA probes.

- *The Spiritual Warrior* by Shakura Rei
  ISBN 1-887472-28-2 $17.95

Eastern philosophy. An exposition of the spiritual techniques and practices of Eastern Philosophy.

- *The Pillar of Celestial Fire* by Robert Cox
  ISBN 1-887472-30-4 $18.95

Eastern philosophy. The ancient cycles of time, the sacred alchemical science and the new golden age.

- *The Tenth Man* by Wei Wu Wei
  ISBN 1-887472-31-2 $15.95

Eastern philosophy. Discourses on Vedanta – the final stroke of enlightenment.

- *Open Secret* by Wei Wu Wei
  ISBN 1-887472-32-0 $14.95

Eastern philosophy. Discourses on Vedanta – the final stroke of enlightenment.

- *All Else is Bondage* by Wei Wu Wei
  ISBN 1-887472-34-7 $16.95

Eastern philosophy. Discourses on Vedanta – the final stroke of enlightenment.

PUBLISHING AND DISTRIBUTING YOUR BOOK IS THIS SIMPLE ...

1. Send us your completed manuscript.
2. We'll review it. Then after acceptance we'll:
   - Register your book with The Library of Congress, Books in Print, and acquire International Standard Book Numbers, including UPC Bar Codes.
   - Design and print your book cover.
   - Format and produce 150 review copies.
   - Deliver review copies (with sales aids) to 20 of the nation's leading distributors and 50 major newspaper, magazine, television and radio book reviewers in the USA and Canada.
   - Organize author interviews and book reviews.
3. Once we have generated pre-orders for 1,000 books, New Author Enterprises will enter into an exclusive publishing contract offering up to 50% profit-sharing terms with the author.

PEOPLE ARE TALKING ABOUT US ...

"I recommend New Author Enterprises to any new author. The start-up cost to publish my book exceeded $20,000 – making it nearly impossible for me to do it on my own. New Author Enterprises' ingenious marketing ideas and their network of distributors allowed me to reach my goals for less than $4,000. Once my book reached the distributors and orders started coming in, New Author Enterprises handled everything – financing, printing, fulfillment, marketing – and I earned more than I could have with any other publisher."

– **Rodney Charles,** *best-selling author of* Every Day A Miracle Happens

PUBLISH IT NOW!

116 North Court St., Fairfield IA 52556 • **(800) 532-4734**
http://www.newagepage.com